EARLY PRAISE FOR *WHAT'S UP DOWN THERE?*

"I laughed, I cried, I wanted to throw my arms around Lissa Rankin and thank her wildly! *What's Up Down There?* is a book of rare honesty, comfort, and humor that casts a feminine eye on the most powerful and vulnerable part of the female body. *What's Up Down There?* has everything I've ever wanted to know about vaginas, but was too shy to ask!"
—SHEILA KELLEY, AUTHOR OF *THE S FACTOR*

"Dr. Lissa has served up one of the hippest, coolest, straight-speaking books on women's genital health, function, and fun. Finally!!!"
—LOU PAGET, BESTSELLING AUTHOR OF *HOW TO BE A GREAT LOVER*

"Every woman needs to be open and honest about her body with her gynecologist. This book answers important questions you may have been afraid or embarrassed to ask. Every woman's body is magnificent and needs to be honored and treated with loving care by her doctor and herself."
—JUDITH ORLOFF, M.D., AUTHOR OF *EMOTIONAL FREEDOM*

"Imagine sitting next to a warm, charming, funny gynecologist on a seven-hour flight where they're handing out free cocktails. Reading *What's Up Down There?* is that sort of experience: delightful, giddy, memorable, and illuminating. What a wonderful book."
—MARY ROACH, *NEW YORK TIMES* BESTSELLING AUTHOR OF *STIFF* AND *BOINK*

"Dr. Lissa Rankin provides answers to the questions every woman always wants to ask her doctor—and some that they've never thought to ask, but should. Best of all, she does it with both humor and style."
—JOHN GRAY, *NEW YORK TIMES* BESTSELLING AUTHOR OF *MEN ARE FROM MARS, WOMEN ARE FROM VENUS*

"Once in a generation there's a book like no other. You're holding it in your hand. Buy it. Bet you'll love it. Bet you'll read it more than once. Bet you laugh and cry each time you do."
—RACHEL NAOMI REMEN, M.D., *NEW YORK TIMES* BESTSELLING AUTHOR OF *KITCHEN TABLE WISDOM*

"Lissa Rankin is a sassy, brilliant, articulate, funny, fun, loving genius. She gets inside the body and soul of every woman, speaks straight from the heart, educating each of us about the privilege of being a woman. She makes me feel calm. Proud. Fascinated. And most of all, *educated* about the magnificent, intricate, breathtakingly beautiful body, woman. Every woman must read this **NOW.** This book will put the power to determine your health back in your hands, where it belongs."
—REGENA THOMASHAUER, AUTHOR OF *MAMA GENA'S SCHOOL OF WOMANLY ARTS*

"Dr. Lissa Rankin has written a courageous book, providing information about topics that few others are willing to tackle. And when she doesn't know the answer, she says so, giving us all the more reason to trust the answers she does give. *What's Up Down There?* dispels myths while reminding us of the beauty and mystery of our girl-bodies."
—DIANA DAFFNER, COAUTHOR OF *TANTRIC SEX FOR BUSY COUPLES*

"Lissa is that approachable Ob/Gyn we all wish we had. In this frank, enlightening, funny book (you know you're in for a treat when the table of contents makes you laugh) she answers all those questions and every other question I've ever had about my girly bits in thirty years of being a woman, twenty years of being a lover (sorry if that's too much information, Mum), and five years of being a mother."
—LORRAINE REGEL, COAUTHOR OF *THE SURVIVAL GUIDE FOR ROOKIE MOMS*

"With humor, honesty, personal candor, and professional expertise, Lissa Rankin breaks through the shame and discomfort most of us feel about discussing the most intimate of subjects. She makes shocking feel comfortable, outrageous becomes commonplace, confusing becomes simpler, and suddenly, discussing menstruation, birth control, genital piercing, and so much more, is like a conversation with a great friend over coffee."
—ELISSA STEIN, COAUTHOR OF *FLOW: THE CULTURAL STORY OF MENSTRUATION*

"Yes, I know *What's Up Down There?* is written for women, but I read it cover to cover, with the kind of ghoulish fascination most men have when contemplating the mysteries of the female body. And it delivered, providing answers to the kinds of questions every guy I know has wondered about, but wouldn't dare ask the women in our lives. Lissa's approach is so straightforward, so funny, and so reassuring that it made me want to dump my own primary care doc and see her instead."
—ARMIN BROTT, *NEW YORK TIMES* BESTSELLING AUTHOR OF *THE EXPECTANT FATHER*

"Lissa Rankin transitions from doctor to BFF without missing a beat, tackling even our most embarrassing questions with the utmost candor and without a speck of judgment. She rips off the flimsy paper sheet we've used to cover any unfounded shame or confusion we've experienced for being female. Embarrassing? Try empowering!"
—JORY DES JARDINS, COFOUNDER OF BLOGHER.COM

What's Up Down There?

QUESTIONS YOU'D ONLY ASK

YOUR GYNECOLOGIST IF

SHE WAS YOUR BEST FRIEND

Lissa Rankin, M.D.

with a Foreword by Christiane Northrup, M.D.

ST. MARTIN'S GRIFFIN 🙦 NEW YORK

This book is not intended to serve as a substitute for professional advice and intervention, and is not intended to replace the advice of a gynecologist or medical professional, who should be consulted about any health care issues that may affect the individual reader.

The information contained in this book is the product of observations made by the author in her practice, as well as her review of relevant literature in her field of expertise. The literature at times reflects conflicting opinions and conclusions. The views expressed herein are the personal views of the author and are not intended to reflect the views of any group or organization with whom the author is affiliated.

All persons described in this book have had identifying characteristics altered in order to protect their privacy.

Note: The descriptive title of this book is not licensed by or associated with any third party.

www.stmartins.com

Book design by Richard Oriolo

ISBN 978-0-312-64436-9

FIRST EDITION: October 2010

10 9 8 7 6 5 4 3 2 1

For my patients, who trusted me with their stories

and allowed me to witness the beauty that lies

within each woman

Contents

Foreword

BY CHRISTIANE NORTHRUP, M.D.

ALL GYNECOLOGISTS SHARE A RARE kind of insider knowledge: up close and personal information about the most intimate details and anatomy of a woman's body. We are privy to women's most well-kept secrets. Secrets that, for centuries, have been shrouded in mystery, silence, and shame. Even today, fifty years after the sexual revolution of the 1960s.

I remember a political cartoon from the 1980s that showed a woman's body with arrows pointing to each part of her anatomy, signifying who that part belonged to. Her brain, her legs, her face, even her breasts were marked: HERS. There was a box covering over the entire pelvic area, however, that said: NOT HERS. Instead, there was a sign posted on that area that read: "Property of politicians, husbands, fundamentalist religious leaders, fathers, feminists, and governments."

Clearly, there is no more highly charged area of our anatomy than our genitals and their functions. Which is why, back in the early 1990s, when I was writing my first book, *Women's Bodies, Women's Wisdom,* I was faced with inventing an entirely new way of describing women's bodies—a positive language that would give women a way to think about and interact with their intimate geography in a liberating and empowering way. I also knew that sharing my own experiences was critical. After all, my generation proclaimed that the "personal is political." And I saw the truth of this daily. Women don't have personal and professional lives. We have whole lives. And what happens to us in the bedroom, boardroom, kitchen, or hospital is all a seamless piece. And all of it sets the stage for our state of health. Any approach that doesn't acknowledge the seamless unity of our bodies, minds, and spirits keeps us out of touch with our power to flourish.

However, back when I was attempting to articulate this new vision of women's health, my brand of holistic thinking was considered heresy. We doctors were supposed to be objective authorities whose recommendations were based strictly on scientific evidence, not anything messy like our own lives or emotions. You were never supposed to share your personal life with a patient. (No one seemed to understand the implications of the Heisenberg uncertainty principle of physics—that the very act of observing a phenomenon changes that phenomenon—and that this principle applies to human interactions, not just particles. In other words, our thoughts, beliefs, and emotions can and do influence our states of health—and the health of everyone around us. Neither doctors nor anyone else can separate the effects of our own thoughts and beliefs on everyone around us. In fact, our own thoughts and beliefs can even be a source of solidarity, comfort, and healing to each other! What a concept!

But that type of thinking was so rare back then that I kept my holistic thinking mostly to myself, certain that my conventional medical colleagues would ride me out on a rail if they found out. Publishing the first edition of *Women's Bodies, Women's Wisdom* back in 1994 was, therefore, an act of faith and courage. (And yes, I'm patting myself on the back here. Big-time. Something I encourage more women to do when they, too, have accomplished something they're proud of.)

Now, a mere fifteen years later, it's a new world. Okay—we women still have a ways to go. Especially in other areas of the world like India, where they still selectively abort about 1 million female fetuses per year. Don't get me started.) But hey, in some regards, we are moving at warp speed. Women only got the right to vote in 1920, a mere blink of an eye in human history, especially when you consider that women and that which we consider feminine have been actively suppressed for the last five thousand years or so.

Given all that, you can understand why it gives me so much pleasure to introduce a new generation of women to the wisdom of their bodies through the work of OB/GYN doctor Lissa Rankin. In the refreshing *What's Up Down There?*, Dr. Rankin freely and honestly shares her own personal and intimate experiences with sex, birth control, fertility, and all things gynecologic. What a breath of fresh air to see that the next generation of gynecologists has picked up the torch that I struggled to light back in the 1980s and '90s.

Dr. Rankin represents the best of the new generation of young women who are honoring the feminine, embracing our female bodies, and doing so with humor, pleasure, honesty, and a sense of fun. Of course Dr. Lissa Rankin is not your ordinary gynecologist even by today's standards.

She is also an artist and a visionary who started a phenomenon called Owning Pink—a community of women dedicated to celebrating all things feminine, successful, sexy, and pleasurable. She's a woman's health visionary—and it takes one to know one! *What's Up Down There?* offers just what every woman needs: a gynecologist girlfriend who answers the kinds of questions that women are still afraid to ask. And she does so with a unique brand of cheeky, girlfriend, hip warmth that tickles me pink. This book is a pleasure. Read it. You'll see.

What's Up Down There?

Introduction:
Let's Talk About
Coochies and Boobs

BRRRRRINGGGGG. . . . **THE PHONE RINGS. CALLER** ID says it's
Chloe. I pick up.

All I hear are giggles. Then a snort, followed by a cackle.
"Chloe?" I say.

Chloe snorts again. I shake my head and smile.

I hear someone yell, "Don't say 'vagina' so loud!"

They've done this before. Chloe and Piper are in Manhattan
celebrating some girl time away from the kids, obviously talking
about sex over a few cocktails. Whenever they think up questions about their girl parts, they call me. They are my best
friends, and I am a gynecologist.

A clanking noise indicates that Chloe must have dropped

the phone. Then I hear Piper's voice. She says, "We've made a bet, and you gotta help us out. If a woman squirts fluid when she has an orgasm, is it pee?"

They collapse into a fit of giggles again, and I wish my job didn't keep me from zipping off to New York on a girl-bonding lark. I want to be with them, leaning on each other for information and support, broaching topics most people don't dare venture near.

I answer the question. Piper wins the bet. We air-kiss into the phone—"mmmwhah!"—and hang up.

Five minutes later, the phone rings again. It's Piper.

"Is it true that your uterus can fall all the way out of your vagina and wind up hanging between your legs?"

I answer the question. This time, Chloe wins the bet.

Ten minutes pass. *Brrrriiing.* Chloe fires off more questions: "Why is it that my crotch smells like fish sometimes?" "Why do boobs get smaller when you breast-feed?" "How come I have a droopy box since I gave birth?"

They put me on speakerphone so that it feels more like I am there, sipping a glass of wine with my girls in a swanky Manhattan bar, instead of sitting at home with my beeper on.

Chloe and Piper like to brag about having a gynecologist on call 24/7. In addition to being their friend, I've delivered their babies, performed their Pap smears, and helped them with issues ranging from postpartum blues to pelvic prolapse. It's a role I'm proud to play.

When Chloe and Piper returned from New York, we sat together with our significant others over shrimp Caesar salads. Chloe, always the rabble-rouser trying to embarrass the guys, asked, "Have you ever heard of a gynecologist finding something weird stuck up someone's vagina?"

So, mid–salad bite, I answered the question. Yup. You can bet I've seen my share of stuff stuck in vaginas. Although actually, the up-the-butt stuff tends to be way more common. ("Really, Doc, I swear, I fell on that Lysol can/gerbil/cucumber/toothpaste tube.") Although there are many colorful answers, one came to mind instantly. The guys at the table rolled their eyes, but everyone put down their forks, riveted.

It was 3 A.M., and Mildred, a frequent flyer to our emergency room, showed up for the umpteenth time complaining of pain in her "passion flower." The emergency room paged me, the gynecologist on call, to come see her. Since I had not met Mildred before, the nurse felt compelled to warn me before I started the pelvic examination. "Just so you know," she said, "Mildred uses her vagina like a purse."

Thinking the nurse was euphemistically informing me that Mildred was a prostitute, I asked how long she'd been hooking. The nurse explained, "No, she's not a prostitute. She literally uses her vagina like a purse. She stuffs it with money, Motrin, keys . . . you know, purse stuff."

I went to see Mildred, who politely shook my hand. "Oh, hi, Doc," she said. "Let me get 'er ready for ya, sweetie." She then proceeded to pull down her pants and begin yanking things out of her vagina like it was Mary Poppins' magic carpet bag (or, in this case, carpet box). There was a plastic Baggie of pills, a wad of bills, a tube of lipstick, a pen. Half-expecting Mildred to pull out a red scarf that magically turns into a bouquet of flowers, I was on the verge of busting out laughing when I suddenly realized that there was something very wrong with this picture.

It's tempting to laugh when gynecologists tell vagina stories. But sitting in that room with Mildred, I realized that something tragic probably had happened that made Mildred think using her vagina as a handbag was a good idea. My heart filled with compassion for her, and when I asked her flat out whether she had a history of sexual abuse, she put her head on my chest and cried like the eight-year-old she was when she was first violated. She admitted that she hated her "passion flower" and figured, since it had done nothing but bring her trouble, she might as well put it to good use. After our early-morning chat about owning and respecting her beautiful, sacred yoni, Mildred swore she would buy a purse and save her passion flower for the purposes God intended. God only knows what happened to her, but my heart still aches to think of her.

With this book, I'm not trying to gross you out or make you lose your appetite for shrimp Caesar salad. Nor is my primary goal to elicit giggles (though you will laugh plenty—as you can see, we gynecologists have some stories!). Instead, I aim to quit skirting the issues, the way many doctors do. I'm not going to tell you the "safe" answer or hedge my bets. I'm not going to worry whether insurance companies will agree with my recommendations or whether lawyers will sue me. I'm just going to talk to you like a friend, someone you can trust to tell it to you straight.

While my girlfriends get to indulge every question they've ever had about coochies, boobs, sex, butts, and women's health—usually giggling with me over a glass of wine—most women don't have a gynecologist at their beck and call and end up talking among themselves, often perpetuating myths and repeating misinformation. This book aims to bridge the chasm between the questions real women have and the answers a gynecologist would give you after she took off her white coat and sat down with you over a cup of coffee (or a cocktail).

To write this book, I solicited questions from friends, family,

Twitter and Facebook buddies, and the community from my Web site OwningPink.com, asking them to send me the questions they would only ask their gynecologist if she was their BFF. The response I got was overwhelming. Thousands of questions came pouring in, most of them with a great deal of overlap. What I discovered was that 90 percent of the questions were variations of "Am I normal?" And 90 percent of the time, the answer was a resounding "YES!" Each chapter in this book begins with my personal story, since I not only roll on a stool between the stirrups, I have also straddled them. The chapter then transitions into Q&A regarding that chapter's topic. By the time you finish this book, you will learn more about me and my vagina than you probably ever wanted to know—the sexual dysfunction I suffered, the sexual transmitted disease I contracted, the puberty moments I endured, and the pregnancy I experienced. But my hope is that you will also learn more about you and your girly parts than you've ever known before, and with that education, wisdom, empowerment, and inspiration, you will more fully love the normal, beautiful, healthy woman I know you are.

I'm going to approach your intimate questions the way I think we should all practice medicine—by practicing love, with a little bit of medicine on the side. What does love have to do with gynecology? *Everything.* Trust me on this. When we approach our bodies with love, acceptance, and nurturing kindness, we pave the way for magic to unfold, the kind of magic I'm blessed to witness every day. So spread your legs and open your heart. Let's explore together what it means to be truly, wholly, and authentically female. You might even discover that being open to the part of you with the capacity to give birth just might help you give birth to *you.*

CHAPTER 1

Being a Gynecologist

ALL WOMEN HAVE STRADDLED THE stirrups, but few have the opportunity to see things from the other side. My patient Lita always brought her boyfriend, who was fascinated with what went on between the stirrups. Lita would undress, lie on the examining table, and mount her feet into the padded pink stirrups. Her boyfriend would hustle over to where I was sitting between her legs on my rolling stool, eye-level with her coochie, and he would beg me to let him take a look.

I always deferred to Lita, who didn't seem bothered by having her boyfriend watch. He would pull up a chair, with his eyes glued to her crotch like he was watching the tie-breaking pass in the last minutes of the Superbowl. I would insert the speculum into her vagina, pull out my cytobrush and spatula to do her Pap smear, and he would sit behind me, resting his chin on my shoulder as I worked. He would pepper me with questions and even asked if he could pull out the speculum when I was done.

It always weirded me out a little bit. I mean, who wants their boyfriend getting all *clinical* on them? Don't you want him to think of sex, not speculums, when he's staring with rapt attention at your hooha? But this guy's fascination with what goes on between the stirrups is not uncommon.

When I was an OB/GYN resident, I briefly dated this babycakes of a boy, who we'll call Adonis, since I honestly can't remember his name. I'm pretty sure he was legal, but only barely, and when he and his buddies found out I was a gynecologist they were riveted. They made up a drinking game they called Gyno Guzzling in my honor. As the star of this game, I was supposed to tell gynecology stories, and every time I said the word *vagina* or *speculum*, they would drink. While the feminist in me found this game a tad offensive, the narcissist in me couldn't resist the attention, and a roomful of hunky college boys egged me on for hours, as I told story after story, until we were all hammered. Gyno Guzzling went something like this:

LISSA: So Jolene comes into the clinic and I'm about to do a Pap smear when the nurse says to me, "Oh, you gotta be careful when you go in there, you know. Jolene's *vagina* has teeth."

BOYS: Drink!

LISSA: So I'm thinking the nurse means she's got chops, like her *vagina's* got bite.

BOYS: Drink!

LISSA: So I go about my business, but when I pull out my *speculum*...

BOYS: Drink!

LISSA: I find Jolene's va jay jay stuffed with those plastic teeth you wear when you're pretending to be a vampire on Halloween.

BOYS: No way!

LISSA: Yes, way!

BOYS: More! More!

LISSA (COYLY, BATTING EYES): Well, okay. So the ER calls and says, "We've got a lady with purple pee. Can you come see her?" So I take a look at the pee and, sure enough, it's as purple as Violet in *Charlie and the Chocolate Factory.*

ADONIS: What's *Charlie and the Chocolate Factory?*

LISSA (SIGHING, THINKING, *I'VE GOTTA DATE OLDER GUYS*): Never mind, sweetie. Anyway, I ask her, "Did you eat anything unusual? Have you had purple pee in the past?" And she shakes her head. I say, "You put anything purple up your *vagina*?"

BOYS: Drink!

LISSA: And she says, "No, ma'am. I ain't." So finally, I stick a *speculum* in there . . .

BOYS: Drink!

LISSA: And her *vagina* is filled with this skanky, gelatinous purple goo.

BOYS: Eeeeeewww. . . . Drink!

LISSA: So I scoop some out with my glove and show it to her, and say, "Then what's this?" And she says, "Oh, that. Well, my mammy told me I should always use jelly when the boys put on them condoms, so I do what my mammy says and use Welch's."

BOYS: Ha Ha Ha Ha Ha . . .

ADONIS (LOOKING MORE GOD-LIKE THAN EVER): One more, Lissa? Please?

LISSA (THINKING, "ANYTHING FOR YOU, GORGEOUS!"):
Okay. So I'm examining this woman with a plastic *speculum*…

BOYS: Drink!

LISSA: And she asks me if she can take home the *speculum* so she can look for her G-spot!

BOYS: Drink!

ADONIS: You must get some great pickup lines when guys find out you're a gynecologist. What's the best one?

LISSA: You mean what's the worst? "Will you be my doctor?"

ADONIS: What do you say?

LISSA (GRINNING): I don't do men. That one always shuts them up.

So if you're curious what it's like to be a gynecologist, you're not alone. When I go to dinner parties with people I don't know well, the conversation invariably winds up going *down there*. Gynecologists evoke a sort of morbid fascination from others, much as morticians must. Like a rubbernecker at a gruesome car wreck on the highway, you can't help being fascinated.

Really, we gynecologists are not so different from accountants, lawyers, car mechanics, or firemen. We all just focus on what we do best. What sets us apart is that what we know best happens to be what makes the world go round.

Why in the world would anyone ever want to be a gynecologist?

I know what you're thinking. We spend all day looking at naked women, so we must be a bunch of perverts, right? Does a woman do it because she's a closeted lesbo? Does a man do it because he's a sex fiend? But it's not like that. I swear. Nobody chooses this field because we get to stare at naked ladies all day long.

My call to medicine came early. I was only seven when I first started nurturing injured baby squirrels back to health, feeding them dog's milk with an eye dropper every two hours throughout long nights. Plus, my father was a doctor, so I grew up in hospitals. By the time I was twelve, I scrubbed in on my first surgery, a hysterectomy performed by one of my dad's friends, who graciously let me, little punk that I was, don a pair of surgical scrubs and stand back with the anesthesiologist with my washed and gloved hands. When the uterus came out, the surgeon handed me the rose-colored chunk of bloodied tissue and said, "The uterus."

I held it in my gloved hands, with a mixed feeling of repulsion and pride, and when we did rounds on the patient later that night I blurted out, "I held your uterus!" She looked at me like I was the fungus on a moldy piece of pizza, but I bragged about it for weeks at school, as if it was something I did every day.

"Oh yeah. I had to do surgery on Friday. Just a uterus. Not heart surgery or anything." So I guess gynecology coursed through my veins early on.

By the time I was a twenty-two-year-old medical student, I decided I would be an ophthalmologist, mostly because every doctor I met tried to talk me out of becoming an OB/GYN. "Terrible hours. No life. Crazy malpractice." The rant rang in my ears. But once I jumped in during my first clinical rotation in labor and delivery, there was no turning back. I became a junkie, and labor and delivery was my drug.

While I loved the adrenaline-pumped hustle of labor and delivery, what attracted me most was that I cared deeply about women and the issues we face. My chosen field is not about vaginas. It's about people. Even after all these years, I sit in awe of the beauty within each woman. Am I vagina obsessed? Well, I'm writing this book right now, so maybe. But really, being a gynecologist is about loving, empowering, and embracing women. Vaginas are secondary. They just happen to need a little TLC from time to time.

Doesn't it gross you out to look at skanky snatches all day long?

No. It's really not like that. When you're a doctor, you get used to dealing with things others consider gross. When I was a medical student, I had a patient with severe abdominal pain who hadn't pooped in ten days. When we did an X-ray, the radiologist unofficially wrote on the wet reading: "Diagnosis: FOS" (doctor speak for "Full Of Shit"). We tried laxatives, enemas—you name it. Nothing worked. So it was my job, as the medical student with the smallest hands, to glove up, slather on the lube, and go poop hunting. There I was, up to my elbow in someone's bum, pulling out one hardened, putrified poop ball after the next. It took hours. Once you've done that, vaginas are a piece of cake.

Truth is, most women primp for the gynecologist. They respect the fact that someone's gonna go face-to-coochie. They shower, trim their pubes, and sometimes even spritz on a little Chanel N° 5. For the most part, I've found that women practice good hygiene, even when I worked in public health clinics with women who were lucky to find fresh water to bathe.

Sure, there are exceptions. My paranoid schizophrenic patient Dalia *never* showered. Every time she came to see me, I had to plug my nose before I could get near her. But skanky snatches are no grosser than the vomit my patients hurl on me when they're in the throes of pain, the loogie a smoker hacks up, or the poop excreted by a woman during childbirth.

If you're a mother, you deal with this stuff all the time. Look at what our kids put us through. But just as you snuggle your little one when she pukes in your hair, we gynecologists do what we must to care for our patients. Sure, sometimes, what we must do is a wee bit distasteful. But just like motherhood, the joys and rewards of helping women outweigh any of the downsides.

What's the kookiest gynecology story you've ever heard?

One of my patients sued me for stealing her labia. Swear to God. I had only been in practice for a few months when Mabel came in complaining of abnormal vaginal bleeding. I needed to examine her, but this was no easy exam. Mabel was about five feet tall and about three hundred pounds, so finding her cervix required advanced gynecology skills. When I examined her, I inserted a small speculum we call "Skinny Minny," then the typical larger speculum, followed by a giant speculum and, finally, the Great Mama of all speculums, "Big Bertha."

Now, mind you, Big Bertha only fits a very small percentage of women. About the length and width of my forearm, this is one hell of a speculum, but the rare woman who needs it has such a long, gaping vagina that you can't even come close to seeing the cervix without it. Mabel was one of those rare women. I counted on Big Bertha to help me see Mabel's cervix so I could biopsy her uterine lining and get to the bottom of why she was bleeding.

I warned Mabel, as I always do, that endometrial biopsies are a bitch. While the cervix isn't particularly sensitive, the uterus protests when you poke something inside, causing a severe cramp, like a labor pain. When I warned Mabel, she looked at me blankly.

I once heard someone talk about gynecology procedures as fixing your car's engine through the tailpipe, and that's how I felt that day. I put Big Bertha inside Mabel's vagina and reached for the tenaculum, a scary-looking two-pronged device used to hold the cervix steady while you do the biopsy. With Big Bertha's help, I managed to identify the cervix and grab it with the tenaculum. This is where we hit a snag.

Just as I clamped the tenaculum onto the cervix, Mabel bellowed, "*Whoaaaaa, Daddy-o, get that fucking thing outta me!*" at the top of her lungs. She pushed against the stirrups like a cowboy on a bucking bronco, standing straight up and knocking out Big Bertha, the tenaculum still clinging to her cervix like a king crab. The tenaculum has a handle like a pair of scissors, and Big Bertha clung to the handle, swinging around between Mabel's legs as she leapt around like a bear with a bullet in its belly. Letting out a bloodcurdling scream, she grabbed me by my hair, kicking at my face, screaming, "*Get it out! Get it out!*" I tried in vain to release the tenaculum as I dodged her bucking bare feet.

I tried to reason with Mabel, but to no avail. I screamed for help, and my nurse ran in just in time to witness Mabel standing on the table, kicking her chubby legs in a flailing gynecologic cancan dance, with instruments hanging from her vagina. My nurse took her hand and, speaking like she was talking to a rabid dog, soothed Mabel back to the table, where I was able to remove the instruments.

Not surprisingly, Mabel never returned to have her biopsy

completed, even when I threw out the word *cancer*. Had she ever come back, I would have taken her straight to the operating room and asked the anesthesiologist to knock her out. However, she never returned my calls or answered the registered letters I mailed.

A few days after our gynecologic fiasco, another doctor called my office about Mabel Nile. He said, "I've got this woman here, Mabel Nile. She says you removed her uterus and her bladder and cut off her labia, with no anesthesia, right there in your office. But I took a look at her, and all her parts appear to be where they're supposed to be. What did you do to her, anyway?"

A few days later, I got a letter from Mabel, addressed to "Dr. Rankinstein." On the outside of the envelope was a child-like drawing of a spiky instrument next to two little rectangular boxes. Written on the envelope in red pen was: "You have something of mine, and I want it back." Inside, I found a note, handwritten on lined notebook paper with scratchy, halting letters: "You stole my labia. Where did you put them? In the lab?"

A few weeks later, I received a notice that Mabel was suing me for stealing her labia. Shortly afterward, the director of security at another hospital called to tell me that Mabel had assaulted one of their doctors and was threatening to hunt me down. Fortunately, she never appeared, and after nearly a year of ridiculous legal hoops my lawyer was able to get the claim dropped. Then, almost three years after I first met Mabel, I received a summons to small claims court. Mabel was suing me again. I had to go to court, sit across from Mabel, and defend myself to the judge.

When I showed up in court, Mabel was already sitting on the other side at the plaintiff's table. I tried to muster every bit of human compassion I could find in my soul. I wanted to open my heart and forgive her for the three years of anxiety and inconvenience she'd inflicted on me. I wanted to forget the letters I had to write to explain the situation to every single insurance company that didn't want me to be a provider because I was being sued. I wanted to let go of the humiliation of the investigation by the California medical board. I wished I could let my heart understand how hard her life must be, living in a paranoid schizophrenic world. I'm ashamed to say I wasn't that big a person. I was pissed, and every cell in my body just wanted to kick her.

The judge said, "Ms. Nile. Please state your case."

"That doctor..." She turned and pointed a sausage finger at me. "*She stole my labia!*" she yelled, slamming her fists on the podium. "She's got 'em in a jar somewhere. In the lab. They're gone. Wanna see?" She started to pull down her plaid pants.

The judge stood up and brought down the gavel. "That's enough, Ms. Nile," she said. Mabel pulled her pants back up.

"*She's holding them hostage!* They're in a jar, somewhere in her lab. I just want my labia! *Tell her to give me back my labia!*" she bellowed. The bailiff stood up beside her, but the judge shook her head. Mabel stared into space, and the judge asked her to take her seat.

It was my turn. "The Superior Court judge has already dismissed the case, your honor. This case has been adjudicated." I then presented my request to countersue for malicious prosecution.

I could see Mabel turning red, like a cartoon character, with steam coming out of her ears.

The judge asked everyone to be seated, and the court was silent for a minute. Then she turned to Mabel and said, "What can we do to make you stop torturing this doctor?"

"*Just make her give me back my labia!*" she cried, throwing a little temper tantrum, beating her fists against the podium and stomping her feet.

The judge shook her head and ruled in my favor. I won my countersuit, and Mabel still owes me one hundred dollars.

That was many years ago, and I have long since forgiven Mabel. I hope she found help, and most of all, I hope she finally discovered that her labia are right there between her legs, where they've been all along.

How do gynecologists (particularly male gynecologists) have a sex life without feeling like they're stuck at the office? How is business coochie any different than personal coochie?

I completely understand this question. When I was young, I lived for Oreo ice cream blended with Reese's cups and Heath bars. But after a summer of working in an ice-cream shop surrounded by all the Oreo ice cream I could eat, not only did my obsession with the frozen dairy treat vanish, but also the mere thought of the stuff repulsed me. All these years later, I still think twice about dipping up a double scoop.

So I can see why you might be curious. Why would a male gynecologist want to go home and eat pussy when he's been looking at it in the stirrups all day long? I'm a heterosexual woman, so I can honestly say my job doesn't negatively affect my sex life in any way. If anything, it makes me feel more comfortable being open and honest about all things vaginal.

But let's turn it around for a minute. Imagine you are a female urologist. Having done a rotation in urology, I can tell you that many of the penises I had the privilege to meet were flabby old socks with pus coming out of them that couldn't get an erection even with the penile pump that had been inserted. But many others were hot, young studly penises, whose hunky owners would drop their drawers, turn beet red, and stare at the ceiling while I poked and prodded. But did it affect my sex life? Not a bit.

I have heard from some doctors that this issue arises. One doctor admits that medical school ruined her sex life, since she can't view her vagina as anything but clinical. But her reaction is rare.

When I get home from work, I'm as hot and bothered by my lover as the next person is. I know it's hard to believe, but to most of us doctors, vaginas (and penises) are simply another body part, not unlike the ear or the mouth. What makes vaginas different from the ear or the mouth is the possibility contained within. In this place of wonder, love is consummated and babies are born. We gynecologists select our chosen field because we get to bear witness to this magic.

Do male gynecologists ever get sexually excited by a patient?

Our egos like to think our sexual organs inspire thoughts of sex in anyone who views them. After all, that's what our pretty pink pussies are supposed to do, right? It's hard to imagine that gynecologists aren't secretly turned on while we work.

I hate to break it to you, but we're not.

Sure, I notice when a stunningly beautiful woman with all the right attributes is on the table. I admire perfect breasts and fabulous figures, just like other women notice sexy strangers on the street or drop-dead gorgeous models in magazines. But once again, it's not sexual. It's as if I'm admiring a piece of art.

What about the hot, hung hunks in the urology clinic? Did I ever feel a sexual stirring when examining their genitals? Nope. Really—honest to God. It just doesn't work that way. I might have the same admiration for a perfect penis as I do for a shapely breast, but it comes from a place of clinical detachment. We doctors see tens of thousands of patients over the course of our medical training, and it comes with the territory to distance ourselves a bit from what we experience. We have to. If we didn't, we couldn't cope with the suffering and loss. I experienced more pain during my medical education than the average student because I am hugely empathic and had trouble attaining the level of detachment many doctors achieve. But the detachment does happen to some degree for all doctors. I promise you, to a gastroenterologist, a rectum is just a rectum. And to a gynecologist, a vagina is just another organ.

But I'm a woman. What about male gynecologists? I asked some of my male colleagues if they ever get sexually excited by a patient, and here's what they had to say.

> **SR:** I've never been turned on by a patient's body or physical characteristics. However, I've met many patients with the most attractive personalities and have developed "mini-crushes" on these patients. When they come in for their annual exam, I'm always excited to see them. But the physical exam is still just that: an exam. The conversations can be very stimulating, but the exam is sterile and boring.

> **JG:** Sure, I notice if a patient is hot. I'm a warm-blooded male, and some of my patients are stunning. In the office,

I have to say I notice breasts more than vaginas. I'm a single, heterosexual guy. From the time I hit puberty, I've been obsessed with breasts, and sometimes I find a patient's breasts arousing. But vulvas and vaginas? No. It's clinical. Unlike breasts, vaginas are much more alike than they are different. It's just part of the job.

AC: The short answer is that I am mentally detached in the office, even if the patient is very beautiful. Sometimes I do Pap smears on my wife, who I find extremely sexually attractive. Yet, even with her, there is no sense of familiarity or intimacy when she's on the exam table. Obviously, my emotions toward my wife are quite different outside the office. For the very same reason, I honestly do not find myself sexually aroused by my patients.

There you go. So if you're worried that you might turn on your gynecologist, don't be. And if you're crushing on your cute gyno, sorry to burst your bubble, but you're better off attracting him the good old-fashioned way, with your wit, your wisdom, your inner beauty, and your loving heart.

What is the most unusual-looking vagina you've ever seen?

I've seen my fair share of unusual vaginas. There's the hermaphrodite baby who had two labia and what looked like a penis where the clitoris should have been. There's the teen girl whose vagina was closed up with an imperforate hymen. There's the post-sex-change transsexual who comes in every year to get a

Pap smear. And there's the Somali woman who had her labia and clitoris cut off and her vulva sewn as part of the ritual we in the West call "female genital mutilation."

But what's the most unusual vagina I've ever seen? It's a tough call. But one patient, Celia, takes the cake. Celia came into labor and delivery, hunched over her giant pregnant belly, huffing and puffing. While she squirmed in the stirrups, I examined her cervix and announced, "She's six centimeters dilated. Take her upstairs and call Anesthesia!"

I ran off to see another patient, but I checked in a few minutes later, and the very experienced charge nurse said, "Doctor, I know you said she was six centimeters, but I examined her after she got her epidural, and her cervix is completely closed. No offense, but I think you made a mistake."

I donned a glove, rechecked Celia's cervix, and said, "She's eight centimeters. Set up the delivery table."

The nurse, looking puzzled, put on a glove and tried once more. Shaking her head, she said, "Dr. Rankin, with all due respect, her cervix is still closed—not a bit dilated."

Suddenly a lightbulb went off in my head. Pulling the headlamp around, I looked more closely at Celia's vagina. Inserting my fingers, I felt first right, then left. As I suspected, Celia had two vaginas, with two cervixes. One cervix was eight centimeters dilated. The other was closed. We were both right.

Celia had not had prenatal care, and I had never met her before. Had I examined her under other circumstances, her condition, which we call "uterine didelphys," might have been picked up earlier. Uterine didelphys falls under a category of congenital birth defects called Mullerian anomalies, which occur when the Mullerian ducts fail to fuse properly during the embryonic stage. Other defects can arise when this process gets disrupted, but uterine didelphys, which usually results in two vaginas, two cervixes, and two uteruses, is the most extreme of the Mullerian anomalies.

When Celia had an ultrasound postpartum, we discovered that, sure enough, she also had two uteruses. One carried the pregnancy. The other remained dormant. So when she went into labor, one side dilated. The other didn't.

When I informed Celia of her condition and explained, she nodded knowingly. Her husband had been trying to tell her for years that having sex with her was like having sex with two different women. Sometimes, she felt as tight as a virgin, while other times, she felt much looser. When I examined her long after childbirth, when her body had returned to normal, sure enough, one vagina was very narrow, and it hurt when I inserted a speculum. The other side, which had been stretched out during childbirth, felt roomy.

When she told her husband, he hugged her and said, "Lucky me! I got two for the price of one. Double the pleasure. Double the fun."

Have you ever heard of a gynecologist finding something weird stuck up a woman's vagina?

A young woman named ChiChi showed up in the emergency room. The chief complaint listed on her chart was "I got vines growin' outta my jojo."

Approaching her, I asked, "What's the matter, ma'am?"

ChiChi stared at me, wide-eyed. "I got vines growin' outta my jojo, sister."

"Yeah, that's what they told me, but what exactly do you mean by that?"

"I mean there's vines coming outta me! Just look," she insisted, pointing between her legs.

I told her I'd examine her after I got the whole story. Impatiently she began to explain. "Well, I was fine, and then a couple days ago, I start feeling something tickling me, and I look down there, and there's vines hangin' outta me."

So I asked the obvious question: "Did you put any vines inside of you?"

"Nope," she answered, looking at me like I was nuts.

I snapped on a pair of gloves and pulled out my speculum, getting ready to examine her in the stirrups. But when she opened her legs, an awful stench filled the emergency room. And sure enough, this lady had vines growin' outta her jojo. I tried to examine her, but I couldn't get inside. The speculum wouldn't budge, as if something was obstructing the vagina.

"Honey, are you sure you didn't put something in here?" I asked, rummaging around for a tenaculum. Clamping onto whatever was obstructing the speculum, I wrangled and pulled, like I was yanking out a baby, when plop! Out came this nasty gray thing with, well, vines hanging off of it.

Clueless as to what ChiChi might have just birthed out her vagina, I asked again, "Are you sure you didn't put something up inside your vagina?"

ChiChi looked at me as if a lightbulb had just gone off and said, "Oh, *that*? It's just a *potato*. MeeMaw told me if I put a potato in there, I wouldn't get pregnant, and wouldn't you know, it worked!"

True story, I swear. Disturbing? Yes. Hilarious? Sure. But there's also an important moral here: Please, *please* educate your daughters about birth control. Abstinence is all well and good (I practiced it until the ripe old age of twenty myself). But if you don't provide your children with accurate sex education, they may wind up in the emergency room with vines growin' out of their jojos. I know I'm preaching to the choir here—you all are reading this book. But my heart went out to ChiChi for all she hadn't been taught. An hour later, she left the emergency room safe and sound, with a birth control prescription and a business card for Planned Parenthood. But I suspect the scars in her psyche took longer to heal.

What's the grossest thing you've ever seen as a gynecologist?

You sure you want to know? Okay, here goes. If you're squeamish, skip this question. Otherwise, you were forewarned. .

Ellie, a gorgeous, well-dressed professional woman, came to my office complaining of vaginal itching and a feeling "like something is moving around inside." I performed the routine examination and noticed a thick white discharge that was suggestive of a yeast infection. I swabbed the discharge with a Q-tip, and slapped some of it on a slide. Noting that the slide cover wouldn't lie flat, I headed to the microscope to figure out what the obstruction was.

Then I saw it: a maggot, slithering on the slide like a wiggly piece of rice.

Now, you don't know me, so you have no idea how repulsed I am by wormy things. I nearly dropped out of medical school when we had to study the worms that crawl under your skin and dart across your eyeballs (another story for another book). But no one had ever prepared me to find maggots in a vagina. Surely I had to be mistaken. I took another look. But there it was—up close and personal. One big, fat, live *maggot*.

Fighting back vomit, I sought counsel from my partners. Certainly, I couldn't just barge into the room and tell this woman she had maggots in her vagina. If someone told me I had maggots in my vagina, they'd have to either anesthetize me or kill me. There's no way I could live in my skin after that kind of news. But I didn't want to lie to her either.

After regaining my composure, I questioned Ellie. Had she put any food up her vagina? Maybe a little funky sex play? Any recent travel to odd places? Swimming in any strange bodies of water? Other sexual oddities I should know about? No. No. No. Ellie said she lived an ordinary life, had a long-term monogamous relationship, worked as an accountant, and never traveled.

After examining her again and shooting water into her vagina with a syringe to "clean her out," I discovered two more maggots. We're not talking piles of maggots here. Just three total. But how in the world did they get there? I have no idea. Ultimately, I told Ellie she had a minor infection (infestation?) and a series of pills should clear it up for her. I also suggested she avoid the temptation to put whipped cream, mayonnaise, or any other food product into her vagina. (You never know what people do in the privacy of their homes.) Two weeks later, after I treated her for, well, worms, a repeat examination revealed that she was free and clear. The itching and sensation of movement resolved. A happy ending to a creepy (no pun intended) story.

Do women spruce up their pubes when they come to see you? Or is it usually a bushy mess down there?

Women can be amazingly shy when it comes to showing me their pubes. I can't tell you how many times I've asked a patient to hop up into the stirrups only to find her crossing her legs under her fluffy robe. When I probe, I discover that Mary (or Jane or Sally or Muffy) is trying to hide her pubic hair. Ladies, I'm a gynecologist. This is what I do all day! I'm a pubic hair expert, so nothing you could show me could possibly shock me. I promise.

My favorite story about pubic grooming is one I heard when a radio DJ invited women to submit their best gynecologist story. I had a feeling I'd heard the winning entry before, so maybe it's urban legend, but it won the contest nonetheless. This woman calls in and says she's been waiting for months to get to see this hot gynecologist who has a mile-long waiting list. One day, his office calls to say there's been a last-minute cancellation. The doctor can fit her in, but only if she can get there in thirty minutes. She's mortified. Usually, she goes through this whole ritual of trimming and grooming and sprucing herself up for gyno appointments, but since she lives twenty minutes from the office, she knows she won't have time. So she goes for the turbo-job, grabbing a washcloth, scrubbing her cooch, and spritzing on some Giorgio before heading out the door.

During her speculum exam, the handsome doctor says, "Well, Julie. You went all out today!"

Assuming he likes the Giorgio, she says, "Well, thank you."

After her appointment, she goes back home and hops in the shower to wash off the gynecology goo. Her eight-year-old daughter suddenly barges into the bathroom, frantically rummaging through all the towels and clothes, saying, "What happened to my washcloth?"

"What washcloth, honey?" the woman asks.

The kid says, "The pink one. The one I've been collecting all my glitter in."

The woman peeks over at the laundry pile, where she sees

the washcloth she used to spruce up for the gynecologist. It is, in fact, pink, and covered with glitter. She looks in the full-length mirror, and sure enough, her vulva is covered with shiny, shimmering sparkles.

Bu dum, ching. So maybe it's urban myth, but it's a classic. Women do get all wigged out about how their pubes look when they go to the gynecologist. The truth is, I really don't give a damn.

You must get a lot of crazy calls in the middle of the night. What's the most memorable?

Well, there was the woman who called me at 4 A.M. to say that she just had the best sex of her life. Then there was the nurse who called to ask if she should wake up Mrs. Jones because she forgot to give her the sleeping pill I had prescribed. Or the 1 A.M. phone call from the woman who just got back from her beach vacation and wanted to know whether her sunburn would hurt the baby. I could go on . . . but here's the most memorable.

"Dr. Rankin, I'm having pain in my hoo-hoo."

"Your hoo-hoo?' I said. "Tell me about the pain."

"Well, it kind of throbs and aches, like it's scratchy and needs someone to itch it."

"Are you scratching it a lot?"

"No. Not really. I can't reach it very well."

I asked all the usual questions: "Are you pregnant? Bleeding? Any discharge?" "Have you tried using any MONISTAT?"

"MONISTAT? Would that help my pain?"

"Maybe," I said. "How long have you had this pain?"

"About six months," she said.

I had been up all night and wanted to yell, "And you're calling me now, at one in the morning?" But instead, I composed myself, took a deep breath, and let out a long sigh. "Sweetie, tell me about your pain."

"Well, when I reach down the back of my throat, there's this little lump that shouldn't be there, and it kind of—"

I interrupted her. "Why are you telling me about your throat?"

"'Cause I've got a pain in my hoo-hoo."

I exhaled, tapping my fingers on my knee. "Is that what you told the operator?"

"Uh-huh."

"Ma'am." I paused, reminding myself to stay calm. "Define 'hoo-hoo.'"

"You know, that little dangly thing at the back of your throat that wiggles around when you sing?"

She had to be kidding. "You're having pain in your throat? Then why are you calling the gynecologist?"

"'Cause I figured my regular doctor would be sleeping by now."

That one I'll never forget.

What's your wildest childbirth story?

Oh my... should I tell you about the woman who came in with no prenatal care and plopped out undiagnosed triplets ("Ma'am, here's your baby. And oh my God! Here's another one. And what's this? A leg? Girlfriend, you're not gonna believe this, but...").

Or the woman who squirted worms out of her butt while pushing her baby out? Or the sexually molested pregnant twelve-year-old schizophrenic, attended by her schizophrenic mother, who were both hallucinating teddy bears on the ceiling the whole time?

Needless to say, I've got a bunch, but one takes the cake. A pregnant woman with major medical problems and multiple abdominal surgeries, including a reconstructed bladder that drained urine into a pouch outside her belly, showed up in labor with her baby presenting butt-first. These days, hardly anybody delivers breech babies vaginally, but back in Chicago, where I trained, more than a decade ago, we didn't automatically do C-sections on every breech baby. And because this woman's belly was riddled with scars, we *really* wanted to avoid doing a C-section. Her labor progressed normally, until right at the end when she was pushing. The baby's butt was right there—you could see it, bluish scrotum swollen and bulging. Then the baby's heart rate dropped to the sixties (about half the normal rate). All of the sudden, we faced a medical emergency.

I was standing there as the junior resident, freaking out. There was no way we were going to get that baby out quickly by C-section without slashing through the woman's bladder and ripping holes in her bowel, but we couldn't just watch the baby die. That's when my attending physician, a total cowboy (a term we doctors use for the fearless among us), jumped in, grabbed a vacuum, stuck it on the baby's butt, and sucked that baby out. All of us—the nurses, the residents, the medical students, and the patient—were in shock. Even the cowboy was visibly shaken. Using a vacuum on a breech baby is an absolute no-no. But he did it, and the baby was just fine. I was so scared I nearly soiled my pants. Makes me tremble just thinking about it.

Can you describe the view from the stirrups when a baby is being born?

Can I just say, "Wowser"? Honestly, watching a woman give birth is the most beautiful thing I've ever seen in my life. Even after ten thousand plus deliveries, I still sit in awe. It's hard to put into words.

First, you start to see the head crowning. The baby's head separates the labia and starts to stretch them, so the whole girly area bulges. As more of the baby's head parts the labia, they spread wider, and sometimes the perineum, between the vagina and anus, begins to split open a bit to make room naturally. I stretch the skin to try to make room and minimize tears, until the baby's head finally pops out. Then, you see a straddled woman with a baby's head sticking fully out of her: a very cool—and very surreal—visual.

At this point, my job becomes more active, as I angle the baby's head first down, and then up, to get the baby's shoulder to deliver, since shoulders of big babies can sometimes get stuck. Once the first shoulder delivers, the second usually glides right out. Then it's a slickery hullabaloo of arms and legs and amniotic fluid and umbilical cord, all coming out together. Oh. And there's blood. Lots of blood. And sometimes pee and poo mixed in.

Then it's clamp-clamp on the umbilical cord, slice the rubbery cord with scissors (usually we give Dad the honor), then wipe with blankets and baby snuggles at Mom's breast. I usually tear up at this point. Gets me every time. Welcome to the world, baby!

CHAPTER 2

How Coochies Look

MY BEST FRIEND FROM HIGH school once asked me, "What's it like to look at vaginas all day long?"

I answered, "Really, it's just a bunch of different haircuts."

Which is pretty much the truth. Having rotated through a urology clinic, I can honestly say that penises vary way more than coochies do. From the other side of the stirrups, we all look pretty much the same. Sure, some women have bushy bright red pubic hair, some have longer or shorter labia, and some have waxed their pubes into a question mark. Some have clitoral piercings and tattoos, while other have moles, warts, or ulcers. But mostly, we're not so different, you, me, and all the rest of us.

So how come the vast majority of questions women submitted for this book revolved around how we look? Why is this so important to us? As women, we are so much more than vaginas, and yet you wouldn't know it from the way our society behaves. Too often, others try to diminish us to nothing more than a piece of tail, but we are oh, so much more than that.

Let me tell you a story that shocked me. I am the founder of a Web site, OwningPink.com, which is all about being authentic, celebrating the various facets of our wholeness and humanity, joining in community, and getting your mojo back. When we launched Owning Pink, readers (we call them Pinkies) showed up in droves, craving the loving, encouraging, motivational words of wisdom we share. Or so we thought....

When our marketing team investigated where most of our traffic came from, they asked me to guess. What Google search landed people at Owning Pink? I'm thinking *authenticity, empowerment, health, creativity, spirituality, women's issues, healing*—something like that. But no. The number one reason people show up at Owning Pink is because they Google-search "pretty pussy" and wind up at the Pretty Pink Pussy Tour, a very benign and unsexy post educating women about their anatomy. This revelation inspired me to post this on the Web site:

> Why are you all out there searching for "pretty pussy" when you could be interacting, loving, creating, even finding yourself some real, live pretty pussy to nurture and adore? Do me—and all the pretty pink pussies on the Internet—a favor. Humanize the pussy, please. Respect the person attached to that yoni, and let that respect overflow into the world at large. Those of us with pretty pink pussies (and beautiful hearts, souls, and spirits to boot) are grateful.

Given that this is the culture in which we live, it's no wonder that the majority of questions women ask me revolve around whether they look normal. We are products of our society and, as much as it breaks my heart, our society values pretty pussy.

Before I answer your questions, let me invite you to reflect for a moment on true beauty. I know it's a cliché, but I honestly believe beauty lies within you. When you tap into the wellspring of your heart and soul, you become radiant—no matter what you look like on the outside. The more you nurture your spirit, spread love in the world, and cherish who you really are, the more beautiful you become, even if your pubes are bushy and your labia are long.

Don't ever forget that you are more than just a vagina. Keep in mind that the vagina is a vessel, a place of possibility, a creative, fertile breeding ground where miracles happen—just like the miracle that is *you*.

Why do we have pubic hair?

It's important to understand that the vagina is one of our most delicate organs. It's like the pearl buried deep beneath slippery, soft oyster muscle, and armored inside a shell bastion. Nature protects the vagina well. Our prehistoric ancestors had hair all over in order to stabilize the body's temperature and protect its delicate skin from the outside world. Although we've lost most of our hair, evolution left us with a patch in the pubic region, which, in addition to keeping our pink pearl warm and debris free, serves several important biological purposes.

For one thing, pubic hair functions as a reproductive billboard to potential mates that you are now biologically (if not emotionally) prepared to procreate. Of course, we thwart biology by covering our precious pubes with designer duds. Perhaps that's why we feel inclined to wear short skirts and plunging necklines when we're teenagers. We're just replacing one billboard with another!

Pubic hair also acts as a pheromone carpet. Pubic hair grows where apocrine glands live (in the pubis and under the arms) and traps the pheromones secreted by those glands. When broken down and mixed with bacteria and sebaceous secretions, pheromones gather in pubic hair and act as erotic aids, attracting potential mates to the promised land. Pheromones also tend to attract partners with specific genetics, promoting biologic diversity for the purpose of creating stronger, more genetically healthy offspring who are able to better ward off disease. Nowadays, pubic hair may seem like more of a nuisance than a biological advantage, but it's there for a reason. Embrace it.

Is pubic hair supposed to be the same color as the hair on your head?

Not necessarily. While in some people, the carpet matches the drapes, so to speak, it's not the case for everyone. The presence of two distinct colors of hair in the same person is called "heterochromia." My patient Molly is a fiery redhead, and she got teased her whole childhood by guys who nicknamed her "Fire Pie," referencing the bright red bush they envisioned. But Molly got the last laugh. Turns out her bush is a dark chocolate brown without so much as a fleck of red. It's not uncommon for people with light-colored hair on their head to have darker pubic hair. Keeps 'em guessing, I suppose.

Why does pubic hair stop growing at a certain length, when the hair on your head keeps growing?

It's not so much that pubic hair stops growing. It just reaches a certain length and then falls out before it has a chance to grow longer. How long your hair grows depends on how much time each individual hair is alive. While it may seem like the hair on your head has been there for twenty years, it hasn't. Whether it's a hair on your head, an eyelash, or a pubic hair, hair lives for a certain period of time; then it falls out and gets replaced. A hair follicle undergoes three distinct growth phases: the anagen phase, when the hair follicle actively produces hair; the catagen phase, when growth stops and the hair rests and then falls out; and the telogen phase, when the follicle rests. How long hair grows depends on how long the anagen, or growth, phase lasts. Those with ponytails they can sit on have hair follicles with very long anagen phases.

While the hair on your head has an anagen phase that lasts approximately three to seven years, pubic hair stops growing after a few weeks. So don't worry if you opt not to groom your pubes—you won't need to braid them any time soon.

Is there a right way to shave my pubic hair? Does shaving my pubic hair make it grow in thicker?

Shaving remains the tried-and-true method of removing pubic hair: simple, cheap, quick, and predictable; we all know we can rid ourselves of unwanted hair in a blink if the razor is nearby. But shaving has its downsides.

We've all been there: It's a hot, sunny day, and the beach beckons. You don your pink polka-dot bikini, only to discover that your Fabulous Furburger is overflowing past the bikini line. Bathing suit still on, you grab the razor and maybe a little hand soap and set to work scraping the pubes off your inner thighs. Problem solved, right? Psych.

Two hours later, you're covered with fire-engine red bumps advertising your bikini shave to the beach-going public. And you can't even swim because you're so raw that the salt water makes you want to rip out your whole genital region and sling it into outer space.

Even in non-emergencies, when you do it right—soak in the bathtub first, lather up your pubes with the best shaving cream, and use a fresh, sharp razor—razor burn, ingrown hairs, itching, and burning often follow.

Why does this happen? In addition to slicing through hair, razors shave off the top layer of the epidermis, resulting in tissue injury. The skin responds as it's supposed to, by increasing blood flow to the area in order to heal the tissue injury, leaving you looking like a sunburned, plucked goose. Then there are the bumps, which result from shaving off and damaging puckered hair follicles.

The easiest way to relieve the red rash of razor burn is to stop shaving and grab some boy-short swimming trunks. However, if the bikini beckons and you're not interested in exposing your Furry Monkey to a beach full of hair gazers, don't fret. There are ways to shave safely and relieve the raging redness.

TIPS FOR SHAVING YOUR PUBES

1. Before you shave, soak in a warm bath, which softens the hairs and allows for a gentler shave. Then exfoliate your bikini region with a loofah or shower puff to remove old, dead skin cells that could clog your pores. Bath products containing salicylic acid or glycolic acid may help prevent ingrown hairs, but some people are very sensitive to chemicals in this delicate region.

2. Purchase high-quality razors and discard them after a few uses. Staph infections resulting from nicks caused by old razors can be a doozy.

3. Choose a shaving cream with aloe or other soothing ingredients, and avoid shaving with bar soap, which further dries the skin, rather than hydrating it. An old-fashioned shaving soap with a badger brush can work wonders. Try leaving the shaving cream on for a few minutes before you shave to further soften the hair.

4. Hold the area to be shaved taut but not stretched. Shave in the direction the hair grows—with the grain, rather than against it. Clean your razor with water between each swipe. If you must double back over an already-shaved area, lather on more shaving cream.

5. Apply ice to your freshly shaved bikini line to hasten the closing of your pores.

6. Try following shaving with products like Tend Skin or Bikini Zone, which you can buy at the drugstore. Aloe vera gel, witch hazel, and tea tree oil sprays offer natural relief.

7. Wait at least thirty minutes after shaving to apply moisturizing lotion. This allows the pores to close after shaving, minimizing irritation.

8. Treat redness and irritation with over-the-counter hydrocortisone cream. This topical steroid cream acts as an anti-inflammatory and constricts raging blood vessels. But save it for serious irritation. Chronic use can thin the skin.

9. If you break out into pimples where you shave, zap zits with benzoyl peroxide cream.

10. Ingrown hairs can lead to pustules. If you are plagued by them, apply warm compresses to the pustules twice a day to encourage the hair to bust through. If this doesn't work and the overlying skin is thin enough that you can see the hair, you can try unroofing the pustule yourself. Clean the area with hydrogen peroxide and use sterile tweezers or a needle to fish the wayward hair out of the pustule. Minimize squeezing and other trauma that can further damage the hair follicle and cause the ingrown hair to recur. When you locate the offending hair, don't pluck it, which can further inflame the area. Just free it and encourage it to grow in the direction of the other hairs. If a pustule is especially large, surrounded by a lot of redness, accompanied by fever, or hurts like the dickens, call your gynecologist or dermatologist right away. We don't want you neglecting those precious pubes.

You may feel like shaving replaces every hair with two or transforms a thin strand of hair into a course, black rope. But the myth that shaving thickens hair growth has no scientific backing. Hair growth and thickness depends on the hair follicle and is unrelated to shaving. Because stubble grows back all at the same length, it may seem darker and coarser. New stubble may come in with sharp tips, making it seem thicker than older hairs that may have softer edges. But these are both illusions—once fully grown, your bush will be the same thickness it was before. So if you prefer to groom your pubes by shaving, shave away.

Is it okay to wax your bikini line at home?

Since I've never seen scientific studies about waxing your bikini line at home and nobody every mentioned it in medical school, I can only comment from my personal experience, so take it with a grain of salt. Most of my life, waxing held no appeal for me. Going to the gynecologist once a year was mortifying enough. (Yes, even gynecologists get uncomfortable in the stirrups.) So baring it all for the waxing gurus to strip away layers of skin cells, stripes of hair, and my remaining modesty for the sake of pubic beauty just wasn't my bag.

But one hot summer day, I decided to try it by myself at home. After all, I'm a gynecologist, I reasoned. How hard could it be? With no one around and bikini season upon me, I pulled out the home waxing kit a friend had given me as a gag gift and plugged in the wax. One simple step and I'd be bikini ready.

Or so I thought.

Hours later, I found myself sitting on one of the blow-up donut pillows we send home with women after they give birth, slathered in aloe vera and numbing myself with an ice pack. Needless to say, I learned a few things.

LESSON #1: *Do not* apply hot wax without testing the temperature first. Hot wax burns the bejesus out of you. And that butt-ugly burn sticks around.

LESSON #2: Let the wax cool completely before pulling it off. Otherwise, the wax will not come off in one fell paper-ripping swoop. Instead, it leaves a gummy hornet's nest of sticky, hairy, tangled goo that scissors can't cut and additional strips won't remove.

LESSON #3: Go out and buy nice skinny wooden applicators, rather than using the humungous two-by-fours they include in the kit. It's impossible to craft a porn star landing strip using a canoe oar. Without the help of a nice, delicate, wooden applicator, you'll wind up with a cue ball for a coochie.

LESSON #4: Don't use the fancy-schmancy scissors you use to cut your bangs to chop out clumps of wax-laden pubic hair. They'll end up in the trash can, stuck to the toilet paper you tried to use to mop up the extra wax.

LESSON #5: Make sure you pee *before* you start waxing. Nothing like acid on a wound to send you through the roof.

LESSON #6: Load up on that wax removal product the kit recommends buying. Since my waxing was a spur-of-the-moment decision, I proceeded without any clean-up aids. Hours later, there I was, pubes tangled in the equivalent of chewing gum, careening bare assed through the kitchen in search of utensils or products that might rescue me from my waxy nightmare.

LESSON #7: Avoid all alcohol while waxing. Halfway through this ordeal, I sought solace from a leftover

margarita, still in the martini shaker from the previous
night's Mexican fiesta dinner party. All this achieved was
a reduction in my inhibitions, resulting in Lesson #8.

LESSON #8: Vegetable oil does not clean up bikini wax.
Sure, it works great to remove the beeswax I use for my
art. But bikini wax plus vegetable oil equals bloody
disaster (literally, by this point).

LESSON #9: Make sure you put your head hair up in a
clip before embarking upon a bikini wax adventure.
When the wax meant for your pubes ends up in your
locks, it gets ugly.

LESSON #10: Think twice about whether you really want
to be a middle-aged woman with the va-jay-jay of an
eight-year-old. I have to say, once all was said and done, I
felt robbed.

And even if I didn't, I'm far too scarred from my one horrify-
ing experience to fly my airplane down that landing strip again,
if you know what I mean. Is it safe to wax your bikini line at
home? Maybe, when forethought and sense are employed. However, the moral of my story is this: Do yourself and your coochie
a favor and seek professional help.

*I know it's normal to have hair above my girl
parts, but what about between my
bum and my vagina?*

Don't worry; you're completely normal. Chances are that your
mother has hair there, too.

The distribution of pubic hair tends to be genetic, and it also varies depending on your ethnic background. In my practice, I've noticed that women from India and Latin America tend to be hairier than most Asians. Women from Mediterranean backgrounds seem to be bushier than those of Scandinavian heritage.

Why do we have differing amounts of hair? An individual's *escutcheon* (the fancy medical term for pubic hair distribution) tends to follow apocrine and sebaceous gland geography. Since it's common to have peri-anal apocrine glands, it's not surprising that you also have hair there. You're not alone—millions of other women do, too.

I'm getting chemo, and I expected to lose the hair on my head. But my pubic hair seems to be falling out, too. What's up down there?

I'm sorry, honey. Sometimes we docs forget to warn people that chemotherapy usually makes *all* hair fall out—not just the curls on your head and the arch of your eyebrows, but yes, you guessed it, your pubes. Chemotherapy drugs attack all rapidly growing cells—whether they're cancer cells, blood cells, or cells in your hair follicles. These drugs are not specific to any particular type of hair follicle cell. The über-strong chemo drugs tend to kill hair cells indiscriminately, leaving you lashless and pubeless. Rest assured, the hair will grow back when you're recovered and moving past this obstacle in your life.

Why is my mountain of Venus so ... puffy?

Wondering why you might be developing what some call a "FUPA" (fat upper pussy area) or "gunt" (cross between gut and cunt)? The mountain of Venus (what we docs call the "mons pubis") consists of pubic hair pointed on top of a fat pad above the pubic bone. Which means that if you gain weight, your mons pubis will likely gain weight, too. If you're a thin woman with a puffy mons pubis, that may just be the way nature created you and your fat pads.

Some women who are very unhappy with their puffy pubic region opt to pursue liposuction. Because the puff comes from fat, it responds to liposuction the same way thighs and bellies do. But what's wrong with a little puff? It's all part of who you are. Nature gave me a fat pad under my belly button that no weight loss or abdominal exercises eliminate. Sometimes we just have to accept our bodies the way they're made.

What's the most common labia size, and please don't say that all vaginas are different and special. Seriously, what's the most common?

I can honestly say I have never pulled out a ruler when a woman is in the stirrups. And like it or not, all vaginas *are* different and special. In general, though, overweight women tend to have bigger labia majora (outer lips) because the fat pads that live in the labia get bigger if you have more fat. Because labia minora (inner lips) do not contain any fat, their size is unaffected by body weight.

But okay, fine. You want numbers, and you're in luck—there's actual data. Gynecologists Bergh and Dickinson must have had a lot of time on their hands, because they *did* pull out the ruler while women were in the stirrups. After examining 2,981 women, they compiled this data about the labia minora:[1]

Size of Labia Minor Based on Examinations of 2,981 Women

LENGTH	NUMBER OF WOMEN	PERCENTAGE
0–¾ inch	2,613	87.7
¾ inch	146	4.9
1¼ inch	170	5.7
1½–2 inches	32	1.1
2–2⅓ inches	20	0.7

If you, like me, find yourself tempted to pull out a tape measure and a hand mirror to see how you size up, let me offer this one suggestion: Lock the door. And put the tape measure back in the tool chest when you're done, before anyone asks you what you've been measuring. Trust me on this one.

My clitoris is really small and my best friend's clitoris is way bigger. What is a normal size for the clitoris?

Like labia size, the size of the clitoris varies widely. But what's average? An article reported the results of clitoral measurements of two hundred consecutive, menstruating women at routine gynecologic examinations.[2] These researchers discovered that the average clitoral glans was 0.13 inches wide and 0.20 inches long.

Dr. Robert Latou Dickinson, author of *Atlas of Human Sex Anatomy*, examined one hundred women and found similar results. In his study, 5 percent of women had a clitoral glans smaller than 0.10 inches, 75 percent measured between 0.10 and 0.26 inches, 20 percent measured between 0.26 and 0.59 inches.[3]

How does that translate? The average glans of the clitoris (the little nubbin that sticks up) is about the size of a BB (a bit smaller than a pencil eraser).

How big is too big? When the clitoris is exposed to extra testosterone, it can grow into a sort of mini-penis, a condition called "clitoromegaly." "What?!" you exclaim. "A woman with a penis?" Well, kinda, sorta. An abnormally enlarged clitoris looks very similar to a penis and can experience similar types of erections. To classify as abnormally large, the clitoris must be about twice the size of an average clitoris. When we gynecologists see a patient with clitoromegaly, it's our cue to go hunting for some underlying medical condition, such as a testosterone-producing tumor or certain adrenal gland disorders.

My inner labia hangs longer than my outer labia, and I feel like Dumbo. Is that normal?

Yes, if your inner labia (the labia minora) are longer than your outer labia (the labia majora), you are 100 percent normal. While some women have inner labia that are enclosed within the lips of the outer labia, others have inner labia that stick out. Robert Latou Dickinson, the sex researcher who went around measuring labia, reports one woman whose two labia minor, when fully stretched, spanned nine inches. Women in certain African cultures would envy her. Some go to elaborate efforts to elongate their labia to appear more beautiful.

So if you were born with long labia, embrace your beauty and diversity. Aren't you glad we're not all clones?

Why doesn't my vagina look like the ones in Playboy?

Can you say "air brushing"? And "Photoshop"? And "surgery"? I can tell you firsthand that most women do not look like the women you see in porn. Seems that porn stars all have neat little labia tightly tucked up between their legs, adorned with clean little landing strips. The skin of their labia is all pink and fleshy, with none of the darker discoloration I often see in real life, and none of them have longer labia. If you pay attention, you might also notice that they also don't have cellulite, moles, belly fat, thick thighs, or surgical scars.

Hmmm…sound suspicious? It should. These women do not in any way reflect how women in my office appear. In real life, we come in all shapes, sizes, and colors, and we are as unique as snowflakes.

Do you recommend labiaplasty or vaginal rejuvenation surgery?

No. On the contrary, I recommend against these surgeries, as does the American College of Obstetricians and Gynecologists. Those who advocate for what some call "designer vaginas" turn me beet red and make steam come out of my ears. I'm a generally loving person, but I think these people, especially the doctors who are out there trying to convince women that there's something wrong with them, should be hog-tied and forced to spend the rest of their days telling women they are gorgeous and perfect just the way they are. We women have enough pressure to appear a certain way: stick skinny with big breasts, small noses and shapely thighs, tight little buns, and thick, shiny hair. The last thing we need is someone suggesting that our coochies are supposed to look a certain way, too. I mean *seriously*, people.

Many women considering these procedures don't understand that the procedures are not without risk. During my training, I was involved in a "vaginal rejuvenation" surgery aimed at trimming up a woman's labia and tightening the vaginal opening. Audrey had always felt self-conscious about the size of her labia, and she felt like her partner used to get more pleasure out of intercourse before she gave birth. With his encouragement, she opted for a little vaginal nip and tuck.

We counseled her regarding the risks of surgery, primarily the risk of scar tissue and painful sex, but Audrey and her husband insisted. After she recovered, she ended up experiencing so much pain at the site of her surgery that she and her husband couldn't have intercourse for almost a year. It took loads of patience and three years of pelvic physical therapy before they could once again enjoy sex.

Vaginal rejuvenation surgeries may also lead to decreased sensation of the clitoris and other genital tissues, potentially interfering with the ability to orgasm. These surgeries also have the same risks as other surgeries—bleeding, infection, damage to surrounding organs, scar tissue, and anesthesia complications. Surgery should never be undertaken lightly.

In my practice, I have only agreed to perform vaginoplasty or labiaplasty on patients who suffer health-related symptoms due to their long labia or loose vagina. In other words, if the labia are so long that they're chafing or causing a rash, or if the woman experiences pelvic prolapse or urinary incontinence, then surgery may be an option. If my patients insist on going under the knife for cosmetic purposes, I support their autonomy but do not choose to do these procedures myself.

Why would anyone ever get a tattoo down there? Does it really hurt?

I asked a woman with a coochie tattoo, and here's what she said:

I have a butterfly tattoo, with the lower wings on my labia and the top wings on my mons. Why did I get it? I guess I wanted a tattoo like no one else's, and I wanted to be able to hide it. I've had my tattoo for about ten years now. I had it touched up about five years ago because it was fading, and I'll need to do it again within the next year. It was particularly painful on my labia but not too bad.

I got a Brazilian wax beforehand so the tattoo artist had a fresh canvas, but you pretty much have to enjoy your tattoo because growing in your hair isn't going to cover the tattoo up. I tease my husband (he likes hair) and tell him that he's turning my butterfly into a moth. I love my tattoo. I've had many partners that think it's hot, but they are really irrelevant. It's a permanent change that I've made to my body. I'm the one that has to live with it for the rest of my life. I think it's beautiful, sexy, womanly, and unique, just like me.

I'm thinking about decorating my cooch. I know tattoos have their risks, but is there anything that makes it riskier to tattoo down there?

As someone with an aversion to needles, I've always had a bit of a hard time understanding why a woman would choose to tattoo her ya ya. Then I met Alexis. She was covered in tattoos and said, "The way I see it, we only get one canvas in life. I like to think of my tattoos as an intricate story of who I am." The canvas metaphor resonated with me as an artist. I was almost tempted to start tattooing my own canvas.

When I was a resident, a frequent flyer to our clinic bore three tattoos. One inner thigh read: "Christmas." The other inner thigh: "New Years." Then tattooed on her waxed mons pubis were the words "Drop in sometime between the holidays," with an arrow pointing straight down between her legs. You gotta give the woman credit for her sassy sense of humor.

So are coochie tattoos safe? For the most part, coochie tattoos are no more dangerous than tattoos elsewhere. As with any tattoo, you must make sure the parlor you select practices universal precautions, since blood-borne diseases can be transmitted via contaminated equipment. And if the tattoo artists are not properly sterilizing tools that have been used on someone else's genitals, you could also be at risk for other types of sexually transmitted infections like warts or herpes.

Most important, if you're thinking about tattooing your cooch, make sure you're doing it for you, and think long and hard about how you choose to adorn your canvas. There might come a time when that festive holiday tattoo doesn't seem so clever anymore, and tattoo removal is no easy process. Don't let anyone pressure you into something that doesn't feel authentic.

My boyfriend thinks I should pierce my clitoris to jazz up our sex life. What do you think? Does it really help? Why do people get pierced?

I think it's completely up to you, honey. Don't let your boyfriend influence you. It's your body—your choice. As far as whether or not genital piercing will jazz up your sex life, the evidence is conflicting. One study surveyed women before and after vertical clitoral hood piercing and found an increase in both desire and arousal after piercing. Another study found an increase in desire but no significant change in orgasm.[4]

As for why people choose to be pierced, Elayne Angel, author of *The Piercing Bible: The Definitive Guide to Safe Body Piercing*, says there are many motivating factors that lead people to get pierced. She says that people often pierce their genitals for sexual reasons. Many do not find their coochies appealing or sexy, so making a deliberate choice to adorn this part of the body may be highly liberating, as well as physically stimulating. Couples may also pierce as a joint venture, getting matching genital piercings as a form of solidarity, meant to ignite the flames of the relationship.

Angel also believes people pierce "to make bold statements about personal freedom and combat the impersonality and pressures of modern life." They might pierce to shun convention, to assert independence, to rebel privately, or to get their mojo back. Some do it...just because. It's a form of self-expression. It's personal and doesn't have to be rational. And I say more power to 'em!

Why do we have hymens?

Hymen, named after Hymen, the Greek goddess of marriage, refers to the membrane that partially covers the vagina. Why does it exist? Before puberty, the vagina is exceedingly delicate. Because estrogen has not yet plumped up the vaginal tissue to make it more resilient, the vagina is very susceptible to injury and infection. During our early years, the hymen exists to protect this fragile structure from foreign bodies, such as wadded toilet paper, peas, and Barbie shoes (yes, all are true stories). After puberty, once the ovaries start producing the estrogen that thickens the vaginal tissue, the hymen serves little functional purpose.

If I find a bump on my vulva, how likely is it that I have cancer?

Very unlikely. While cancer can attack the vulva, it happens relatively infrequently. Vulvar cancer mostly afflicts women after menopause and is more common among smokers and those with HPV infection. Only 5 percent of female genital tract cancers occur on the vulva (endometrial, ovarian, and cervical cancer rack up the majority of reproductive cancers). Each year, there are 3,460 new cases of vulvar cancer, as opposed to 40,100 cases of endometrial cancer, 21,650 cases of ovarian cancer, and 11,070 cases of cervical cancer in the United States.[5]

So if you find a bump on your vulva, bring it to the attention of your doctor, but don't freak out. Chances are good that it's an infected hair follicle, a mole, a wart, a skin tag, or some other bump that won't threaten your life.

I've heard that, in some cultures, women get their labia and clitoris cut off and sewn shut. Is this true? If so, how can they have sex and babies?

Yes, sadly, this is all too true. I had the opportunity to work in a public health clinic where most of my patients were refugees from Somalia, and every single one of them had their genitals mutilated as a rite of passage during childhood. In Somalia, as well as other African cultures, prepubertal girls gather with other girls and women for a ceremony that includes cutting off the labia and the clitoris. A matchstick-sized hole is left for urine to come out, and the rest of the vulva is sewn closed. While her wounds heal, the young girl is attended by others who have experienced female genital mutilation (in their case, FGM Type 4) as part of a coming-of-age ceremony. Once she survives this process, the girl is considered a woman and can attend the ceremonies of other girls getting FGM.

For the most part, my patients were quick to defend their cultural rituals. I tried to be culturally sensitive and not impose judgments upon them, but from a health and feminist perspective, I found it impossible not to consider their rituals barbaric. Fortunately, FGM is illegal in the United States; but it still happens underground. As their doctor, all I could do was educate women and pray that they would spare their daughters the scars inflicted upon them.

Can women with FGM still have sex and babies? Yes. But it's not pretty.

Somaya, an unusually empowered Somali woman with FGM, was about to get married and asked me if I would surgically open her up before her wedding night.

"I don't want to get bruises on my head," she said.

I was confused. Bruises on her vulva I could visualize, but bruises on her head? "Bruises?" I asked.

"Yes," Somaya said. "That's how they open you up on your wedding night. They push against you with their penis and bang you against a wall until they rip you open. I don't want to get ripped open, and I don't want those bruises on my head." She said it very matter-of-factly, with no evident emotion. I, on the other hand, wound up in tears.

As long as a woman's vagina is closed (with the exception of a tiny hole barely big enough for egress of urine), she can prove to her husband and his family that she is still a virgin—a requirement for any bride entering into an arranged marriage. When she loses her virginity, he forcibly opens her, tearing open the scar tissue.

Once her vagina is opened, the vagina usually stays open, as long as she continues to have sex while the ripped-open wound heals. (I can't help crossing my legs when I think about how much pain these women must endure.) Once the vagina is open, the woman can conceive, and the tearing that usually results from childbirth opens the vagina further.

As you can imagine, complications from FGM abound. Many of these women—the lucky ones who don't die from complications of the often unsterile FGM procedure—wind up with chronic urinary tract infections, painful intercourse, the inability to orgasm, and fistulas (holes that connect the vagina and the bladder, and the urethra and the vagina, or—most distressingly— between the rectum and the vagina; women with fistulas usually suffer from involuntary loss of pee and/or poop from the vagina). Complications of FGM also extend to the woman's babies. Childbirth can result in fetal injury or death resulting from obstructed labor.

Can women with FGM still feel pleasure without a clitoris? Apparently, some can, probably to the chagrin of those who seek to oppress their sexuality. Some woman described experiencing vaginal orgasms, and one woman admitted that she still experiences pleasure by masturbating over the scar where her clitoris used to be. Most likely, this occurs because the clitoris runs deeper than the nubbin we can see, and while the glans of the clitoris may be gone, the body of the clitoris still exists. Or perhaps the capacity for human adaptation exceeds even our wildest imaginings.

How do they turn a vagina into a penis for a sex change operation?

To turn a woman's body into a man's body, many surgeries are required. You must remove the breasts, take away the ovaries, and reconstruct girly bits into a manly member. To do this, a penis is created either by giving hormones to grow the clitoris into a small penis or by inserting an inflatable prosthetic that can be pumped into a penis on demand. Either way, the urethra can be rerouted so you can stand at a urinal and pee like a dude. High doses of testosterone alter your physical appearance, giving you a more masculine form and facial hair distribution. So when all is said and done—if everything goes as planned—you can pee, get it on, and feel comfortable in the skin you're in. Don't we all want that?

I think a lot of people misunderstand transsexuals, labeling them as strange, freakish, or immoral. It's easy to make fun, criticize, or pass judgment on those who are different, and transsexuals certainly make an easy target. But imagine how hard it would be if you looked in the mirror and saw the wrong gender. People going through this difficult process need our love, not our judgments. I'm all about living authentically, and if it takes gender reassignment surgery to let your essential self shine through, I say, "You go, sister!" (I mean "brother.") Whatever it takes to inhabit a body that better fits your sense of self.

If a guy gets a sex change and turns into a woman, does she need Pap smears?

Not technically. A transsexual man who gets a sex change to turn into a woman does not have a cervix and cannot get cervical cancer, so Pap smears are not necessary. But let me tell you a little story about Shania and what I learned from her about gender identity.

Shania used to be Shane. For many years, she felt like she was living in the wrong body. When her brothers expected her to play with Star Wars paraphernalia, she preferred dressing up like Sandy from *Grease* and belting out "Summer Nights" in a poodle skirt. In junior high, when the Sadie Hawkins dance rolled around, she wanted to invite Donny, but her brother shamed her into skipping the dance. When she was home alone, she'd strip off her button-down shirts, thick leather belts, and khaki pants and dress up in her mother's panties, dresses, and high heels. Only then did she feel like herself.

Not until she moved to the big city did she discover she was not alone. Others who looked like men felt like women on the inside. Some even had surgery to rectify the error in nature that kept them from feeling authentic and whole. Shane saw a doctor and began saving pennies. Five years and tens of thousands of dollars later, Shane became Shania, inside and out.

Every year on the dot, Shania took time off from her job and showed up at my office for a Pap smear. I prescribed Shania's estrogen therapy, which helped her look and feel more feminine, but that's not the only reason she came to see me. Every year, she requested a pelvic examination and a Pap smear.

The first time, I found myself dumbstruck. I read her chart, which said: "Genetic Male—XY chromosomes." I'm a gynecologist—I don't do men—so I had no idea why this patient was scheduled to see me. I asked flat out, and Shania simply answered, "Dr. Rankin, I'm here for my Pap smear. It's been a year."

I flipped through her chart and, sure enough, I found a Pap smear report from exactly one year ago, to the date. I read last year's Pap smear. The pathologist reported: "No endocervical or ectocervical cells detected. No pathologic findings." In other words, there was no cervix to Pap, so no cervical cancer was detected.

I simply didn't understand why Shania wanted to waste her money on a Pap smear when she didn't have a cervix and couldn't acquire cervical cancer.

I said, "Shania, you don't need a Pap smear. You don't have a cervix."

She said, "I know, but I want one anyway. That's what we women do."

After doing her Pap smear for the third year in a row, I couldn't contain my curiosity and asked Shania why she wanted a Pap smear every year.

Shania said, "Dr. Lissa, every time I walk into your office, I see other women sitting around the waiting room, flipping through *Cosmo*, holding babies, putting on lipstick. I see all these women—pregnant and breast-feeding and doing lady things. I see the girls behind the front desk, laughing and whispering to each other and talking about their weekends. I know I will never give birth or breast-feed or be quite like the other women in your waiting room, and these doubts about who I am haunt me sometimes. I wonder whether I am really a woman or whether I'm just pretending, like I was as a kid dressing in my mother's clothes. Sometimes, I suspect everyone can see right through me, and that they're laughing. And I get very sad, because inside, I know I am a woman. I always have been. I try to say, 'To hell with the rest of them,' but deep down, it still hurts. So I get unsteady sometimes, wondering." She took a breath and continued.

"Then I come here, to see my gynecologist, and I get a Pap smear, just like all the other women of the world." She reached out and held my hand. "And that's when I know I'm really a full-blooded, honest-to-God woman." She smiled a crooked grin. "No man would ever be caught dead in a gynecologist's stirrups."

I learned a lot that day from Shania. We all do what it takes in order to feel whole.

Do hermaphrodites really exist or is that some urban legend?

Yes, hermaphrodites are more than just urban legend. The historical term *hermaphrodite,* derived from joining the names of Greek god Hermes and goddess Aphrodite, refers to an individual with both male and female reproductive organs. In the mid-twentieth century, the term *intersex* largely replaced the term *hermaphrodite* within the medical community. Recently, the Intersex Society of North America, along with others in the activist and medical community, have moved to replace *intersex* with *disorders of sex development* (DSD).

Regardless of what you call it, this is a real issue. I've delivered quite a few intersex babies myself. The baby slides out, the nurse wipes the baby off, and the excited mother cries, "Is it a boy or a girl?"

And then there is dead silence in the room. The nurse taps

me on the shoulder, so I let go of the umbilical cord, leaving the placenta still inside the womb, to sidle over to the baby warmer, where the baby lies spread-eagle under bright lights. The nurse eyes me, that silent question mark we in the medical field perfect to express confusion or concern without alarming our patients. I see that the baby has two labia, and protruding between them is something that looks like a penis. Is it a boy or a girl? We can't tell.

Communicating this to a parent is not easy. There's no good way to say, "Your baby appears to have what we call 'ambiguous genitalia,' meaning that we see characteristics of both male and female genitalia. We will need to do a blood test to determine your baby's chromosomes, so we can sort out what's going on."

As you can imagine, this does not go over well. We live in a black-and-white world. You're either male or you're female, right? Well, not always.

How do we deal with this? Helping the family determine how to raise an intersex child requires a team of professionals trained to handle this sensitive subject. Historically, gender was usually decided quickly, based more on the appearance of the genitalia than the chromosomes, and these babies underwent surgery as infants. Because it's easier to trim down a penis-like organ than it is to enlongate it, these babies were often quickly and indiscriminately assigned female gender, even if they had male chromosomes.

We've learned a few things since then. These days, there is still a push toward assigning gender, but delaying surgery until the child can be involved in the decision. We've learned that while surgery can address the appearance of the genitalia, this does not necessarily take into account the complex psychological and behavioral factors that influence gender identity.

If you're tempted to make fun of hermaphodites, please resist the tempation. It's hard to feel different. Individuals who are born with this congenital issue need your compassion, not your judgment.

I've heard of vagina dentata, but I'm not sure if this is just a myth. Do some women really have teeth in their vaginas?

Vagina dentata, Latin for "toothed vagina," refers to the folktales told in some cultures about women whose vaginas have teeth that can bite off a penis. These tales, told as cautionary tales to discourage certain sexual behaviors, such as rape, prey on castration anxiety and the archetypal fear that a man may be swallowed and diminished by a woman. Vaginal teeth also symbolize the dark side of feminine power. If a vagina can have teeth, you'd better get out of her way, because she can eat you alive.

But is it real? Do vaginas ever have teeth? They can. I've certainly never seen a whole set of functional chompers ready to lop off my finger during a pelvic exam, but it is possible for a vagina to have teeth. On rare occasions, dermoid cysts develop from pleuripotential cells, cells that may grow into any kind of tissue in the body. These cysts, which most commonly occur on the ovary but can sprout up elsewhere, may contain hair, brain, thyroid, skin, and, yes, teeth. If a dermoid cyst on the ovary ruptures, these teeth may migrate their way through the vagina, resulting in—you guessed it—vagina dentata. But do these teeth have jaws that can bite you? Nah. That part is pure mythology.

I get lost when I look down there. Can you help me find my way around?

Absolutely, and thanks for asking. You'd be surprised how many women go through life never really knowing what's up down there. Did you know we have three different holes? Loads of people don't. One of my highly educated friends recently told me she always assumed that we peed out of our vaginas. It's a logical assumption. After all, men pee and ejaculate through the same pipe. It's only natural to think women would also pee and have sex via the same organ. But nope. Three different holes—the pee hole, the vagina, and the poo hole.

Humor me and let me take you on what I call the Pretty Pink Pussy Tour. (I can see you blushing already. "Oh my, that doctor just said 'pussy'!") Don't be shy or embarrassed. *It's your body.* You have a right to know what's what.

WHAT YOU'LL NEED FOR THE PRETTY PINK PUSSY TOUR

1. A private room with a door you can lock

2. A hand mirror, or a full-length mirror and a lot of flexibility

3. A nonjudgmental mind

4. A smile on your face (yes, giggling is highly encouraged)

TAKE THE TOUR

STEP 1: If you're one of those limber yoginis, just straddle up to a full-length mirror and open your legs all the way, so you can get a good look at yourself. If you're

not that limber, lie on your back, frog-legged, and hold
the hand mirror so that you can see yourself.

STEP 2: Take a gander at yourself, and release all
judgment. If you hear yourself saying, "Ewww...how
ugly," try turning your negative thoughts into
affirmations, such as, "Thank you, Yoni, for all of the
pleasure you bring me." Make a commitment to knowing
and loving your body, just as it is.

STEP 3: Approaching your body with this sense of
gratitude, let's begin.

THE VULVA

First, let me get you oriented. When you look in the mirror,
you're going to see a mound of pubic hair at the top—this is
called the "mons pubis." It doesn't serve much function other
than alerting your sexual partner that there are some hidden
gems underneath your bush (more on this later). When you spread
your legs apart, you will see your vulva: the whole collection of
outside parts. Within it, you will notice two sets of lips, or *labia*.
The *labia majora* consist of the two meatier, outermost hairy
lips. Just inside the labia majora are the *labia minora,* the two
thinner, non-hairy inner lips. These outer structures serve to
protect the sensitive structures that lie beneath the surface.

THE CLITORIS

When you spread the labia open, you will see the rest of
your genitals. If you look just below the mons pubis, the first
thing you'll come across is your *clitoris*. Much of the clitoris is
buried where you can't see it, and the *clitoral hood,* the flap of
tissue that partially covers the *glans* of the clitoris, leaves only a

very small portion of the clitoris exposed. The clitoris is the nerve-laden nub of tissue at the very top of your genitals, just below the mons pubis. This is the only organ in either the male or the female body designed exclusively for sexual pleasure. Wow! Good thinking, God. Way to take care of us girls.

THE URETHRA

If you go down from the clitoris, the next major landmark you'll hit is the first of your three holes—the *urethra,* which is the tube that connects your bladder to the outside world. You urinate out of your urethra, and your paraurethral glands (or Skene's glands), located just inside the urethra, are believed to result in the elusive and controversial female ejaculation. (In some women, during some orgasms, fluid may be expelled from the urethra. More about that later.)

THE HYMEN

Moving further south, you will come across the opening to your second hole—the *vagina,* which is a much larger hole than the urethra and serves several important functions. The *hymen,* or what remains of it, lives right at the entrance to the vagina, right at the *introitus.* If you imagine the vagina as the sleeve of a man's dress shirt, the hymen is its cuff. Usually, in adult women who have had sex, the hymen looks like a raggy, pink, fleshy circle around the vaginal opening, which may have several breaks in the circle or no longer be visible at all, especially if you've had children.

THE VAGINA

Just past the hymen is the vagina, which is a potential space, meaning that if nothing is holding it open, it collapses on itself like a sock without a foot in it. But the walls of the vagina are stretchy, allowing it to expand. When you look at it, you won't see a giant cavity but rather an opening that can expand to serve its function. This is the mother of all pussy places. The vagina is the place where sexual intercourse happens, and it also serves as the birth canal, stretching to allow a baby to come through. Outside of reproduction, it is the place where menstrual blood leaves the body, where the controversial G-spot lives, and where any number of pretty pink sexual activities take place.

THE PERINEUM

As you head south after you leave the vagina, the next thing you encounter is the *perineum,* the tissue between the vaginal opening and the *anus,* the opening to the rectum. The perineum is where you might see an episiotomy scar or healed laceration if you've delivered a baby vaginally. It is also the most common area infected by certain sexually transmitted infections, such as herpes and genital warts. Functionally, the perineum serves to separate the vagina from the rectum, with all its potentially harmful fecal bacteria, but recreationally, this very sensitive tissue is part of sex play for many couples.

THE ANUS

Last but not least comes your third hole, the anus, leading to your *rectum,* which is at the tail end of the gastrointestinal tract that begins at your mouth. Surrounded by the *anal sphincter,* a muscle that serves to hold in poo and gas, the anus looks like a mouth that just ate a lemon, all puckered up and wrinkled. Like the vagina, the rectum is a potential space, so when there's nothing in it, it collapses in on itself, but when it's filled with feces, it dilates, and the anal sphincter relaxes to let it out.

STEP 4: Pat yourself on the pussy! Thank yourself for taking the time to know your body better and to affirm your girly parts for all the beautiful things they do for you. To truly love yourself, you must love all of you. You can't just hide it under panties and skinny jeans and pretend it's not there. It's all part of being a woman, and learning to be comfortable in your skin.

CHAPTER 3

How Coochies Smell and Taste

MANY WOMEN I MEET ABSOLUTELY despise their vaginas, as if they completely buy into whatever childhood messages they were fed about how the vagina is "dirty" and "bad." For these women, any odor wafting up from down there acts as a big stinky banner of how much they hate their girlness. With vagina nicknames such as "fish taco," "crotch mackerel," "cod canal," "fish factory," "fuzzy lap flounder," "tuna town," and "raw oyster," it's no wonder we worry about how we smell. But I say it's time to change all that. Why should we hate what's normal, healthy, and part of the rich female experience?

One of the most common questions people ask me regarding what it's like to be a gynecologist is, "Doesn't it stink?" They wrinkle their noses, furrow their brows, and raise eyebrows, as if there's something wrong with me for loving my job. Lying underneath that question I often see something that borders on misogyny, as if women are nothing more than a vaginal odor to be avoided. From the time we're children, we're taught that normal bodily functions are "yucky." Pee, poop, and privates all elicit a "pee-yew," so it's no wonder we grow up obsessed with how we smell.

Ladies, vaginas are *supposed* to smell. Let me quote my heroine, Eve Ensler, the Queen of Vaginas, whose *Vagina Monologues* has done as much for the vagina as Martin Luther King Jr. did for civil rights:

> My vagina doesn't need to be cleaned up. It smells good already. Don't try to decorate. Don't believe him when he tells you it smells like rose petals when it's supposed to smell like pussy. That's what they're doing—trying to clean it up, make it smell like bathroom spray or a garden. All those douche sprays—floral, berry, rain—I don't want my pussy to smell like rain. All cleaned up like washing a fish after you cook it. I want to taste the fish. That's why I ordered it.

Amen, sister. It's supposed to smell like pussy.

Sure, hygiene plays a role, and just like washing your pits and your feet, cleaning yourself down there is part of being an accepted member of society (not to mention being a conscientious sexual partner). Most women even shower, shave, and primp a bit before visiting the gynecologist. I often notice wafts of perfume emanating from the nether regions. I appreciate the respect and notice the effort, but really, it's not necessary. We gynecologists are not as sensitive as you might imagine.

So how is the vagina supposed to smell? It depends. When

you're straight out of the shower, your coochie may have no smell at all. When you've just finished running a marathon, it may have a strong musky odor from all the sweat glands. When you're menstruating or giving birth, the flinty-iron smell of blood prevails. When yeast overgrows in the vagina, you may smell like freshly baked bread or a good malt beer. Right after you've had intercourse, you may smell faintly bleach-like, as semen has a classic odor of its own. And when certain normal bacteria overgrow, they release amines that smell—yup, you guessed it—like fish.

Every vagina has its own special smell—a combination of the normal bacteria that live in your vagina, what you eat, how you dress, your level of hygiene, your bowel habits, how much you sweat, and what your glands secrete. Remember that the glands near the vagina also secrete pheromones, meant to attract a sexual partner. So you don't want to deodorize your vajay-jay so much that it smells like rain. Doing so thwarts the primal function of what your smell is supposed to accomplish. Plus, it interferes with the vagina's natural pH balance and can lead to a whole host of gynecologic conditions.

So own your odor, girlfriends. Sure, if you're worried, see a gynecologist to make sure your vagina is healthy and normal. But as long as everything's kosher down there, accept that your coochie smells exactly how it's supposed to smell.

My crotch gets extra-funky sometimes. Not to quote a douche commercial, but why do I have that not-so-fresh feeling down there?

Even once you've accepted that your vagina is supposed to smell like you, you might wonder why it gets extra-funky from time to time. When I was in college, a friend of mine got labeled "Tina the Tuna," because a few guys who went down on her told their fraternity brothers she smelled like fish. What made Tina tuna-like? Chances are she had *bacterial vaginosis,* a common type of vaginal infection.

Bacterial vaginosis (BV) causes the classic fishy smell and is caused by an imbalance in your vagina's natural bacteria. In a nutshell, the good bacteria decrease and the bad bacteria over-grow. This causes sloughing of the epithelial cells in the vagina, which may result in a profuse vaginal discharge, as well as the release of smelly amines that result in a fishy smell.

Other causes of funky crotch include poor hygiene and ex-cessive sweating. After a few days of not showering, you'll likely start to reek. Working out can also increase the funk, because your girly parts are full of sweat glands, just like your armpits. Some women naturally have more sweat glands on the vulva, and more sweat can lead to more odor.

If you feel like your crotch is funkier than usual or your partner mentions that you smell suspiciously like halibut, check in with your gynecologist. Chances are, your problem can be treated.

Should I douche? If so, how often?

No, ladies. Please don't douche. Whoever decided douching was a good idea must have hated women. Elissa Stein, who coau-thored *Flow: The Cultural Story of Menstruation* with Susan Kim, says, "For years, they sold Lysol, the same bottle as the bathroom kitchen germ killer, as a douche. They launched a hor-rendous scare tactic ad campaign that assured women their hus-bands would leave them if they weren't fresh and clean. Not only that, women believed Lysol was a spermicide and douching af-ter sex could prevent or end pregnancy." I mean, seriously, people. Lysol? In the poor innocent cooch?

Women who douche keep me in business, but unless you're looking to fill up your gynecologist's schedule, don't do it. The vagina is a self-cleaning organism. Shoving it full of things meant to make you smell like a bouquet of flowers does more harm than good. Douching washes out the vagina's normal bac-teria, allowing bad bacteria to overpopulate the delicate environ-ment and increasing the risk of vaginal infections. Some people can't even tolerate using soap or bath gel on their private parts, since it can lead to itching, burning, and vaginal infections. Be-lieve it or not, warm water on a soft washcloth is all you need to keep yourself clean.

My crotch sweats like crazy and soaks my pants. What can I do about it? Do they make crotch antiperspirant?

Your crotch and butt crack are loaded with sweat glands, and some people have more active sweat glands than others. If you're soaking your pants during spin class or boot camp, you're not alone. Most women's workout clothes get wet when they work up a good sweat. But if you're soaking your pants while sitting in an air-conditioned office, you may have a condition called "hyperhi-drosis," which is characterized by excessive perspiration.

I used to date a guy whose forehead would break into a sweat so frequently and profusely that he coated his forehead with antiperspirant, especially before big meetings. But what about your crotch? Does antiperspirant or deodorant help?

While I'm not aware of anything marketed specifically as a crotch antiperspirant, there are things you can do to reduce wetness and odor caused by excessive sweating. Try washing this area with plain glycerine soap. While some dermatologists recommend antibacterial soaps, such as Dial, pHisoHex, or Lever 2000, these soaps contain harsh products and may even contain harmful endocrine-disrupting chemicals. After making sure you're clean, try blow-drying your crotch. Then apply over-the-counter Zeasorb powder to absorb excess moisture. If you are overweight and have thick skin folds in your groin, try placing clean handkerchiefs between the folds to absorb excess moisture.

If you've tried all this and excessive sweating is still hampering your lifestyle, see a doctor. Your problem may be related to a medical condition or a side effect of a drug you're taking. If your condition is severe, your doctor may discuss treatment options that are generally used for hyperhidrosis, such as antiperspirants with a high concentration of aluminum chloride, which can be applied at bedtime to the labia and groin. If all else fails, talk to your doctor about whether BOTOX treatments or prescription medications may help you.

Why do guys say vaginas smell good when most of us think we smell bad?

Can you say "pheromones"? Pheromones are chemical signals that trigger a response in your mate, signaling biologically that you are available for breeding. These pheromones communicate powerful signals through your natural smell. Why do guys think you smell good down there? Why are they perfectly happy to plant their faces right in your cooch? Because you smell like *sex* to them! We are conditioned to think that certain smells are "good" and other smells are "bad," but the truth is that every individual finds different smells appealing. In fact, studies show that when presented with smelly shirts worn by different men, women preferred the sweaty scent of men whose HLA types (a part of our genetic makeup) were compatible with their own.[1] This means that smell may be more than just a sexual signal—it may even help us choose partners whose genetic makeup combines with ours to produce hearty offspring. So if a guy says he likes your smell, trust his nose.

Aside from douching, are there natural things you can do to make your vagina smell more fresh?

I'm going to assume that you're bathing daily and that you've seen a doctor to make sure you are infection free. But is there anything else you can do to improve your scent naturally?

1. Eliminate hair from your pubic region. Because hair traps odors, any funk you carry down there will get trapped in your pubic hair until you bathe again. Just think about how your hair smells when you've been in a smoky bar. Bare skin does not trap odors the same way hair does.

2. Wipe with baby wipes instead of toilet paper. If funky urine or residual fecal bacteria contributes to your odor, this will help.

3. Drink cranberry juice, which acts as a natural

antibacterial in your urinary tract and may affect your smell down there.

4. Go panty free whenever you can to keep the cooch aired out. Choose cotton panties when you do have to cover up.

5. Take probiotics to keep the vaginal flora healthy.

6. Avoid panty hose and tight jeans.

7. Keep in mind that things that make your urine stinky can lead to vaginal odor. Eat a healthy, largely vegetable-based, whole-foods diet, and avoid known offenders such as coffee, asparagus, beets, alcohol, broccoli, onions, garlic, and curry. Try drinking lots of green juice and eating raw foods instead. Drink lots of water to keep hydrated and dilute the urine.

Is it safe to put perfume on your coochie to make it smell pretty?

While spraying yourself with Jean Naté or Giorgio may work for some women, others will wind up with skin irritation, vaginal infections, and breakouts. Because perfumes and colognes may contain harsh chemicals and drying alcohol, contact dermatitis symptoms are not uncommon. If you insist on trying to make your cooch smell like the cologne counter at a department store, try spraying scent on your inner thighs or buttocks, where the skin is less sensitive.

My vagina smells funny, and I think I might have lost something up there. What should I do?

Though I've found many weird things inside vaginas, what I find most often are ordinary things gone astray. Things like condoms, tampons, and diaphragms mysteriously disappear into the proverbial Bermuda Triangle, never to be seen again. I can't tell you how many patients have called, freaking out because they can't find their condom/tampon/NuvaRing/diaphragm/pessary, et cetera. And yes, if something has been left up there for a while, it usually reeks.

Here's the take-home vagina lesson I feel called to impart. Let's call it the "first law of gynecology," which states that *the vagina does not connect to the lung*. This means that if you lost a condom up there, it's not going to wind up in your circulatory system. Your tampon won't float up through the spinal canal and land in your gray matter. The vagina is pretty much a pouch. Imagine a sock with nothing but a toothpick-sized hole at the end of it. If a condom gets lost in a sock, it's gonna stay in the sock. Also, no connection exists between your vagina and your abdominal cavity that's big enough to let anything but semen and other liquids through. So first and foremost, don't panic. Yes, it's important not to leave things lingering in the vagina indefinitely because of the risk of infection. But chances are, if you think you lost something inside, either it's sitting within finger's reach or it was never there in the first place.

If you think you've lost something in the vagina, go hunting for it, but please be gentle. Don't be afraid to straddle one leg up on a chair, lube up your fingers with K-Y Jelly or olive oil, and stick them all the way inside. If you hit a hard bump that feels like your nose with a little hole in it, you've probably reached your cervix, located at the very end of the vagina. The most common place for a foreign object to lodge is in the crevices on either side of the cervix, so root around a little, and explore both sides. If you can't feel anything, chances are good that there's nothng there and you simply forgot that you already removed that tampon (or diaphragm or condom or whatever). Maybe it fell out into the toilet or landed on the sheets or slipped out on its own. God only knows.

If you're sure something is in there and you can't get it out, ask your partner to go hunting for you. Maybe someone with a better angle can find it for you. If you're too embarrassed or your partner can't find it either, call your handy dandy gynecologist. And don't be ashamed or humiliated. We see this *all the time*. Really. Truly. We don't mind. It's what we signed up for. Vaginas are our job, and I promise we don't judge you for it.

Eighty percent of the time, when I fit a patient into my schedule who thinks something is stuck in her vagina, I insert a speculum and find nothing. On the flip side, it's not uncommon to see a patient who comes in complaining of a foul smell who winds up having something stuck up there (often unbeknownst to her). I mourn the loss of my patient the minute I have to tell her there's a tampon crammed way up near her cervix. When I ask her how long the tampon (or condom or diaphragm or whatever) may have been up there, she visibly cowers, shrinking farther and farther down until she is lying on the exam table in a fetal position, hoping she just evaporates. To make matters worse, I have usually removed the offending object at this point and the whole exam room stinks to the high heavens. We both have to plug our noses to finish talking. Due to sheer embarrassment, many of these patients never come back to see me again. I chalk it up to the Bermuda Triangle Syndrome.

While toxic shock syndrome is always a risk when a foreign body gets stuck inside, most of the time, removing the foreign body—and sometimes applying a little antibiotic cream—fixes the problem and the vagina heals itself. But the wounds inflicted upon the fragile psyche apparently take longer to mend.

If something is stuck and embarrassment stops you from calling the gynecologist, promise me this: Do not, I repeat: *Do not* go hunting with pliers. Did you hear me on that one? One of my patients ripped herself to shreds by hunting for a lost tampon with pliers from her husband's toolbox. Instead of grabbing the tampon (which wasn't even in there), she grabbed her cervix. It took us hours in the operating room to put her back together again.

On Sex and the City, *they talk about guys having funky spunk. Can girls have funky spunk, too?*

God, I miss *Sex and the City*! I think the episode to which you refer finds Samantha grossed out by the taste of her lover's semen. Comparing it to asparagus gone bad mixed with Clorox, Samantha finds herself on a mission to sweeten up her lover's spunk. But what about women? Do we all taste the same, or do some of us taste funky? That's purely a matter of opinion. Just like I prefer Oreo ice cream and you might prefer Butter Brickle, what tastes good differs from person to person.

What does a vagina taste like? I'm afraid the answer to this question lies well outside my job description. Having never tasted a vagina, I glanced around to make sure no one was looking, stuck my finger inside myself, and licked my finger. My answer? My vagina tastes something like a mixture of plain yogurt, bleach, and kiwi (TMI?). Aside from my unscientific self-study, I had to ask around.

When I queried my husband, he looked a little bashful and said, "You're gonna write about this in your book, aren't you?" I nodded. All he'll confess is that I taste much sweeter now that I drink five green juices/day. "Less musky," he says. Since hubby wasn't getting very specific, I had to ask my male, lesbian, and bisexual friends. Here are the answers to my fill-in-the-blank question, "The vagina tastes like ____":

1. "Ripe mangoes"

2. "Licking a battery"

3. "A copper penny, especially around her period"

4. "Seawater"

5. "Asparagus, feta, tuna salad with artichokes"

6. "Lemon mixed with baking soda"

7. "Salty with a tang of soap"

8. "Turkey and applesauce"

How is a woman supposed to taste? The pH (or acid/base balance) of the vagina hovers around 4, the same pH as a glass of wine. Does the vagina taste like pinot noir? Not exactly. But some things that change the pH of the vagina, such as vaginal infections, douching, soap, and exposure to semen, may change the way you taste. Curious what you taste like? Check yourself out. The more you discover about yourself, the more empowered you will be.

My boyfriend says my vagina does not taste good. Are there foods that will help improve the flavor of my vagina?

If you're as sensitive about your smell and taste as most women, I apologize on behalf of your lover. No woman wants to hear that she doesn't taste yummy, but I give him two thumbs-up for honesty. I can only assume that he's motivated to improve your taste so he can give you some love, if you know what I mean.

Is there a way to make the vagina taste better? Maybe. As you can imagine, there's not a lot of scientific evidence to support a link between diet and vaginal taste. (I just cracked myself up trying to imagine getting *that* study past the Institutional Review Board!) But some swear that certain diets and habits can change the taste of your vagina. Rumor has it that cigarettes, marijuana, and alcohol can make you taste bitter, while red meat, dairy, garlic, onion, curry, broccoli, asparagus, spinach, and multivitamins may make you taste salty or sharp. Vegetables high in chlorophyll, such as wheatgrass, parsley, and celery, are reputed to make you taste sweeter, as are spices like cinnamon, peppermint, and cardamom. Eating fruit may also sweeten your taste, as can drinking lots of water.

Vegan diets, particularly diets high in raw vegetables and fruits, are reputed to make you taste fresh and sweet. My hope is that regardless of how you taste, your lover will be singing the praises of the delicious mouthful that is *you*. A note to your partner: Bon appétit!

What should I do if my partner doesn't like to go down on me?

To quote one of my girlfriends, "Tell him to get the hell out. Real men love to eat cunt." (Okay, so my friend tends to say what she thinks without much of a filter.) Seriously, though. If you enjoy oral sex and your partner hesitates, open up the conversation. Here are some tried-and-true tips from Dr. Michelle Gannon, San Francisco psychologist, couples therapist, and founder of Marriage Prep 101 workshops:

1. Ask your partner why you're not getting any. Try to listen to your partner's personal preferences without being defensive or hurt. Some prefer going down on a woman if she is freshly showered, while others prefer less hair down there. We hear more often that those who are reluctant to give oral sex don't like their partner's smell or taste. Some are reluctant to suggest things that might help. "It would hurt her feelings if I ask her to take a shower first." We challenge them that it probably hurts her feelings more that she's not getting oral sex!

2. Your partner may avoid oral sex due to feelings of uncertainty about what to do down there. Time for some fun lessons like those taught at stores like San Francisco's Good Vibrations or Marin's Pleasures of the Heart. Try slipping your partner a copy of Lou Paget's book *How to Give Her Absolute Pleasure: Totally Explicit Techniques Every Woman Wants Her Man to Know.* A little knowledge goes a long way.

3. Some guys have been raised to believe that oral sex is dirty or wrong, when really it can be a healthy, loving part of a relationship. If your partner has hang-ups, consider a few sessions of couples therapy.

4. Try flavored lubricants—mint, chocolate, vanilla, oh my!

5. If you're hoping your partner spends more time down there, make sure to communicate that you really like it. Your partner needs positive feedback, too. Also be willing to return the favor.

6. Face it. Some partners are just selfish and don't want to fully focus on a woman's pleasure. If this describes your partner, discuss your feelings and explain how it feels unfair and nonreciprocal. You want an egalitarian relationship both in and out of bed.

CHAPTER 4

Sex and Masturbation

WHEN PEOPLE FIND OUT YOU'RE a gynecologist, they automatically assume you are a sex goddess who's got it all figured out. What they don't realize is that, just because we memorized the blood supply to the clitoris and the nerve pathways of the pelvis doesn't mean we know anything about sex. Sure, we hear stories. Women talk to us and ask questions, so we gather wisdom through experience. But when it comes to our personal lives, most gynecologists I know are just as messed up about sex as everyone else. Why? Because sex has nothing to do with arteries, neurons, or skin folds.

I never thought much about the true nature of sex when I was younger. Like most kids, I learned that the penis goes inside the vagina and sex makes babies. Later, I learned from peer pressure that sex keeps your man happy and prevents him from abandoning you. Somewhere along the way, my mother told me that sex is a manifestation of marital love, a physical embodiment of the human connection bound by the sacrament of marriage. The combined message left me feeling a little lost when the time came to explore my own sexuality.

The media certainly didn't help matters. The romantic comedies I adore led me to assume that losing my virginity would be an earth-shattering journey to nirvana. But in reality, I suffered a painful, awkward, pleasureless cherry popping in the arms of my college boyfriend—we'll call him Don Juan. While I loved him and relinquished my virginity willingly, it was mostly just a bone I threw him to reward him for good behavior—an alternative, more than anything, to having him dry hump the leg of my jeans in his dorm room while his roommate was at a frat party,

In my mind's eye, I can still see it like it was yesterday. There I was—a twenty-year-old virgin (nearly an old maid, according to the other girls in my dorm). Sweeter-than-gelato Don Juan was trying so hard to make it fun for me, but what did he know? In a hurried embrace of a one-night stand he'd lost his virginity to a local girl on the beach of the tropical island where he grew up.

After locking the dorm room door and clumsily undressing, we climbed into the tiny twin bed where we had already spent many sex-free nights spooning. I was already on the Pill for gynecologic reasons and must have skipped the sexually transmitted disease chapter in my biology books, so the idea of condoms didn't even occur to us. (Shame on you! Bad Lissa.)

We kissed, but not in the passion-soaked way we had in the past. Instead, we kissed timidly, with awkward tongues and bumbling lips. Our noses kept bumping and our hands didn't know what to do, but since we had committed to making it happen that night, we forged ahead. We fumbled with jean zippers and tangled panties until suddenly, with a rush of blinding pain, the moment was upon us. I couldn't think straight—not due to ecstasy but rather from searing, razor-sharp agony. I screamed, and Don Juan quickly pulled out, horrified to be hurting me. But I encouraged him to go on. The poor guy had waited a *long* time for this. It was the least I could do. After a few more stabs of the knife, I felt him quiver, followed by a sudden burning inside me. And then, apparently, it was all over.

After my first sexual experience, I seriously pondered a lifetime of celibacy, but I liked men too much. Embarrassed to discuss my issues with anyone, even my closest girlfriends, I clammed up and resigned myself to a lifetime of martyrdom for the sake of love.

For a decade, sex hurt like the dickens, but wanting to please my boyfriends, I shopped for sexy lingerie, popped champagne corks by firelight, and went through the motions of being a good lover. Sadly, I did this at the expense of my authentic self. Deep inside, I wanted to scream, "*Stop!* Stop the madness! Why are we doing this? I'm hurting and you're having fun. It's not fair, and I'm pissed." But I quieted the voice of my truth and learned coping mechanisms instead.

I would essentially leave my body every time I saw a penis coming my way. Part of me would run for the hills, screaming bloody murder, leaving my partner making love to a cerebrum, but that cerebrum learned to moan at all the right times and follow a pattern of acceptable sexual behavior, which was enough to keep relationships afloat for a few years but certainly not enough to blow anyone's mind.

The fact that I was training to be a gynecologist only high-lighted the irony of my sexual disappointment. All day I gazed at vaginas, answered questions about intercourse, and preached sexual wellness. Others sought me out as a guru, as if I had it all figured out. And my ego preferred it that way. It instilled me with a sense of worth that balanced out the worthlessness I felt in the bedroom. Back then, I never told anyone the truth. I felt like a fucking hypocrite.

Not until the searing pain disappeared in my thirties did I begin to imagine what all the fuss was about. When I left my unhappy marriage and fell in love again, the searing pain I had experienced for a decade magically disappeared. Imagine that. (I tell you, the coochie is wise. Anyone who says there isn't a mind/body connection is nuts.) Not until then did I begin to imagine what all the fuss was about.

As the veil of pain lifted, a tiny part of my essential self be-gan to emerge, one sexy baby step at a time. But it takes time to reclaim your body when you've sent her to time-out for a decade. Slowly, I began talking with my girlfriends about the details of my two failed marriages and sharing stories of my sexual dys-function with patients experiencing similar issues. Telling the truth liberated me and began to shake loose some of the cob-webs of my sexuality.

Now that I've just entered my forties, I feel a shift happening, and it excites me. While society seems to associate sexuality with youth and beauty, I believe we more truly embody the full rich-ness of our sexuality as we age. By shedding the façade of who we think others expect us to be, we more clearly step into who we re-ally are, and with that, the potential for true sexual bliss awakens. I feel like I'm just beginning to walk this path, exploring how two people might connect, not just physically, but spiritually.

For me, part of this requires slowing down in the rest of my life to prioritize a sexual connection with my lover. It's easy to get so consumed by family life, ambition, and the details of daily existence that we think of great sex as just the icing on the cake—something nonessential and merely decorative. But I'd argue that sex is one of the essential ingredients in life's cake—just like a healthy body, an outlet for creative expression, a strong sense of self, loving relationships, a balance between meaningful work and child-like playfulness, and a spiritual path. If you skip the flour, the eggs, the baking soda, or the sugar, you'll wind up with a soupy pile of unleavened batter, instead of the light, airy, delicate, enriching, whole pastry you know you can be.

One thing I've learned about sex is that one size never fits all. We must each walk the path to sexual awakening in our own way, knowing that we walk it in good company. In this chapter, I will try to answer your questions with some generalizations that might help, but keep in mind that no one formula works for everyone. If anything I write doesn't resonate with you, take it with a grain of salt and look deep within yourself. Chances are, the answers you seek have been there all along.

I don't even know what turns me on. How can I get in touch with that?

I hear you, sister. How can you possibly choose between mint chocolate chip and Butter Brickle if you don't even know what ice cream is? Regena Thomashauer, Sister Goddess, author, and founder of Mama Gena's School of Womanly Arts, says:

Most women don't know what turns us on. How could we? It's not like our moms pulled us onto their knees and said, "Puberty is coming, so we're gonna learn what turns you on." And you can't leave it up to your boyfriend. Guys don't know what turns on a woman.

That's why I write and teach what I do—to educate women about how to begin to learn about their relationship to pleasure. If you don't know what pleasures you, you'll never get in touch with your desires. Take the time to learn the difference between what it feels like to touch the palm of your hand, how it feels to run your fingers across your belly or down the inside of your thigh. What parts of your pussy feel good? What pressures do you enjoy? Without learning, how can you allow your lover to gratify you?

There's a scene in the Julia Roberts movie *Runaway Bride* where someone asks her what kind of eggs she likes, but she doesn't know. When she dated a guy who liked scrambled eggs, she ate scrambled eggs. When he liked fried eggs, she ate hers fried. When he liked hard-boiled, she ate hard-boiled. In one scene, she finally lines them up and tastes them all, so she can make a decision, independent of any man.

A woman can definitely be seduced into running that kind of experiment with her own body. Taste. Touch. Experiment. Discover.

My boyfriend wants to know how to please a woman in bed, but talking about sex makes me so uncomfortable. Do you have any suggestions for how to bring up touchy subjects?

I asked for guidance from Lou Paget, a certified sex educator and bestselling author of *The Great Lover Playbook: 365 Sexual*

Tips and Techniques to Keep the Fires Burning All Year Long, who recommends—first and foremost—that this conversation takes place while you're vertical, not horizontal. It's hard enough to talk about sex without doing it in the middle of lovemaking. Talking about it while you're sitting over a cup of coffee takes the judgment out of it.

SOME TIPS FROM LOU PAGET

1. If you wish to give guidance during sex, limit your directions to just one word—*left, up, harder, down.* A sentence is often heard as criticism. A word is heard as direction.

2. If you feel comfortable, try masturbating in front of your partner. Your boyfriend is a guy who is open to listening. Even the world's best athletes have coaches. They need someone who can show them and guide them.

3. Don't *ever* say you want to try something you did with someone else or you wish your boyfriend would do it like some other guy did. If you do, he'll be thinking about that other guy the whole time he's trying to pleasure you. People can't help comparing themselves.

4. Try spicing things up. Your sexuality is an appetite. Sometimes you want a bowl of soup, sometimes you want a full meal, and sometimes you just want dessert. You need to know how to feed those appetites.

5. Don't practice psychic sex, thinking your partner should know what you want. It's crazy making. Communicate what you need.

Remember, sex is a dance. Be happy he's asking for tango lessons!

I used to be so horny, but not anymore. How come my get-up-and-go got up and went?

I hear you, sister. We've all had those moments when blaming a fake headache seems like the best way out. There's no single answer to your question. A woman's libido is as multifaceted and changeable as a kaleidoscope. Do you feel good about your body? Are you living in your skin, or are you completely out of touch with what your body feels? Are your hormones in balance? Do you have medical conditions that might affect blood flow to your genitals, such as high blood pressure or diabetes? Are you taking medications that affect sex drive, such as antidepressants or birth control pills? Do you feel nurtured, loved, and safe in your relationship? Are you prioritizing sex or just adding it to your to-do list at the end of a busy day? Is your partner making time for foreplay before going from zero to sixty straight into your vagina? Have you been sexually abused or raped in the past? Do you feel so stressed out that you can't relax? Are you getting enough sleep? All of these factors and many more influence how horny you feel.

You may even notice that your sex drive comes and goes like the weather. One day you feel like a dog in heat, and the next day you don't give a flip if you never have sex again. Many women notice cyclic changes during their menstrual cycle. Some feel most randy during their menstrual cycles, while others notice spikes in their libido during ovulation. Suffice it to say that we're all different and a woman's sex drive defies complete understanding. If you're feeling turned on and your partner's up for it, enjoy those times. And when you wish your partner would take a hike, express your feelings, but be sensitive. Maybe your off day is the day your partner is dying to jump your bones! A sexual relationship requires compromises to achieve long-term bliss.

Where can I find my little lost sex drive and how can I get it back?

A woman's libido can be a slippery little sucker. Unlike men with decreased libidos, who often respond to a little vitamin V (Viagra), women's needs go deeper. After ignoring women's sexual needs for decades, pharmaceutical companies are finally paying attention, trying to make big bucks marketing a magic pill to turn us on. So far, studies show that no single thing gets us in the mood. While drawing blood flow to the penis usually does the trick to turn on a man, increasing blood flow to the clitoris or vagina doesn't necessarily do the trick in women. Our needs tend to be more complex.

SOME TIPS I RECOMMEND TO MY PATIENTS

1. **SCHEDULE SEX DATES.** If you're up late doing the dishes, doing the laundry, and doing the children's homework, you won't feel like doing your lover. Plan ahead and *make* it happen.

2. **FAKE IT TILL YOU MAKE IT.** Not that you should ever fake orgasm (I highly recommend against doing so), but going through the motions of being sexual can get your juices flowing, even when you're not in the mood.

3. **BUY THE BOOK** *101 NIGHTS OF GRREAT SEX: SECRET SEALED SEDUCTIONS FOR FUN-LOVING COUPLES,* **BY LAURA CORN—AND USE IT.** It's filled with tear-out pages of sexy seduction scenes "For Him" or "For Her." It works every time, if you're daring enough to be a little naughty.

4. **EXPERIMENT WITH EROTIC MOVIES, BOOKS, OR MAGAZINES.** Send your inner critic to time-out and see how you feel. Keep an open mind.

5. **INVITE SOME SEX TOYS INTO THE BEDROOM.** If you usually keep your vibrator in the bottom drawer, under a pile of magazines, yank it out and welcome it into playtime with your partner. If you've never experimented with sex toys before, consider going shopping.

6. **SEE YOUR DOCTOR.** A battery of tests can look for reversible causes that can be addressed. You might also consider a trial of hormone replacement therapy. While studies to support its use are limited, some patients report improvement in their libido when using low-dose testosterone replacement. If your estrogen and progesterone levels are low, as they are in menopause and may also be premenopausally, you may benefit from replacing these hormones as well. Talk to your doctor about the risks and benefits if you're interested in exploring how hormones might help your sex life.

If I take some of my husband's Viagra, will it jazz me up for sex?

I know it's tempting, the way those little blue pills beckon from the nightstand while he's brushing his teeth in anticipation of making love to you. He's raring to go, but you're fantasizing about curling up with a good book. What if you just snuck one of his Viagra? What would happen?

Unfortunately, not much. While studies of Viagra in women have shown some promise in improving the diminished orgasmic function of women on antidepressants, Viagra fails to get women horny. While Viagra is known to bring blood flow to the clitoris the way it brings blood to the penis in men, women seem largely indifferent to this physical change. Why? Because arousal in women relies on many complex factors, which are usually more psychological than physical.

Sex always hurts me. Why does everyone else think it's so much fun?

Oh, sweetie, I'm sorry. Sex is supposed to be fun, right? If your only exposure to sex was through watching movies, you'd think all women love sex, everyone orgasms with intercourse alone, and sex never hurts. You would be very wrong. My practice and my workshops are full of women who experience pain or fail to experience pleasure during intercourse. As I mentioned earlier, in my twenties I was one of these women. I suffered from a now-healed condition that made having sex feel like I was getting stabbed with a knife while someone poured acid inside me.

Painful sex (we docs call it "dyspareunia") is normal when you first lose your virginity. But after a while, the pain should lessen and then resolve. If it doesn't, a variety of conditions can be responsible, including *vulvar vestibulitis* (inflammation of the vestibule), *vaginismus* (involuntary contraction of the vaginal muscles), allergic reactions to things such as latex condoms or spermicide, and *endometriosis* (when lining from the uterus gets on the ovaries, bowel, and pelvic lining). As women age, they also have lower levels of estrogen, which can make the vagina thinner, leading to *atrophic vaginitis*. Many of these conditions can be treated, but it's important to know the reason for your pain. If you find yourself having sex regularly and are still having pain, be sure to tell your gynecologist. There may be a simple solution.

Feeling safe in a relationship is also an important part of making sure sex feels good. Those who have had traumatic experiences with sex in the past, such as being the victim of a rape or child molestation, often experience pain with intercourse. Women's bodies can be amazingly powerful at manifesting physical signals in response to psychological issues. If any of these issues might play a role in the pain you feel, please see a therapist who can help you work through what you've experienced.

In the meantime, make sure you are adequately aroused before intercourse. Encourage your partner to assist you with a little foreplay. If that's not enough, try a personal lubricant. Sometimes, lubing up can decrease the friction against the vaginal wall. Also, certain positions may be more comfortable than others, so experiment with what feels better and listen to your body. If all else fails, get creative about ways to satisfy each other sexually. Oral sex, sex toys, and hand jobs just might be your best friends until these issues resolve.

Most important, be honest with your partner about your experience. Don't pretend you're having fun when you're not. You don't want to train your body to endure pain when it's supposed to be experiencing loving connection and sexual pleasure. Doing so can teach your body to shut down, which can damage your ability to be properly aroused in the future. Get help, be honest, and know that this won't last forever.

My daughter masturbates regularly. I'm secretly worried she's going to grow up to be a sex maniac. Is she normal, or is my child a pervert?

It's perfectly normal for children to masturbate. Just as they inspect the ladybugs they find on the sidewalk and experiment with how they can make fart sounds with their hands, they're exploring their bodies. Children masturbate because they're curious, it feels good, and it soothes them. So rest assured, your child is not a fledgling sex freak.

While you may be tempted to speak harshly when you see your child with her hand in her pants, try to avoid the temptation. How you react to her behavior imprints upon her young psyche and may influence both her current and future feelings about her body and her sexuality. If you overreact, you signal to your child that there's something wrong with what she's doing, a sentiment that may affect her sexual relationships later in life. The best thing you can do if you catch your child masturbating in the privacy of her bedroom is ignore it. If your child insists on touching herself in public places, gently teach her that this is something private that should be restricted to the bedroom or bathroom at home. Use this opportunity to help build the foundation for a healthy sexual life in your daughter's future.

I know I'm not gonna get hairy palms or anything, but can masturbating too much hurt me in other ways?

Not usually. Most women suffer no ill effects from masturbating. On the contrary, masturbating can provide pleasure, diminish sexual frustration, aid in sleep, relieve anxiety, benefit your health, and enhance your sexual relationship with your partner. While some argue that excessive masturbation makes you feel less sensitive, making it difficult to orgasm with a partner, others swear that masturbation keeps them primed and spices up their partnered sex lives. Of course, if you're getting off in the middle of a board meeting, while performing surgery, in the midst of testifying to a jury, or in some other inappropriate venue, you might have a problem. But assuming you're in the privacy of a quiet room, it's all good.

One notable exception comes to mind. Sophie decided to get it on with a banana. Shortly after beginning her sensual ritual, Sophie noticed a growing wetness and found a scarlet circle of blood on her bed. She called an ambulance, and when I examined her in the emergency room I found that the banana, which she had inserted stem side in, had lacerated her cervix. A few stitches did the trick, but I can't help but think Sophie still feels a little scarred.

Of course, Sophie's problem wasn't that she decided to partake in a little self-love fest. It was more the manner by which she went about it. The moral of this story: Have at it with masturbation, but if you're going to stick something inside, stick to nice smooth fingers, vibrators, and dildos and save the bananas for your cereal.

Do you have a favorite vibrator? I want to try one, but there are too many choices when I look at sex toy catalogues. Can you help me choose?

Every woman is so different that it's almost impossible to answer this question. After Charlotte refused to leave her bedroom after acquiring the Rabbit on *Sex in the City*, many women—myself included—discovered the ecstasies offered by this particular pink pleasurer, which gives the "twice is nice" clitoral stimulation and a small amount of penetration all in one toy. But this is not the magic bullet for every woman.

Chrystal Bougon, founder of Bliss Connection, says, "I feel strongly that it's important to find the right toy for each woman. As much as I love the rabbit-style toys myself, I don't think they are a good fit for every woman. I like to start women who are new to sex toys with a vibrator that is focused on clitoral stimulation alone. For most of us, this is where the action lies. Keep in mind that intercourse is not designed with the female orgasm in mind. It's designed for reproduction, and the clitoris sometimes gets neglected during lovemaking because it's hard to reach during partner play. For those of us who can only achieve an orgasm from clitoral stimulation, a bullet-style toy like the sweet, simple, and affordable Silver Bullet is almost a guaranteed orgasm. Once women are more comfortable with the clitoral toys, I then graduate them to the Rabbit Habit or the Original Rabbit Pearl you mentioned (also known as 'Charlotte's favorite' from that now-infamous episode of *Sex and the City*.)"

The key is to find what works *for you*.

How come nobody talks about the wet spot? Is it all sperm, or is there other stuff involved?

I hear you on this one. When I lost my virginity, the semen-soaked wet spot really shocked me. It doesn't seem fair that he gets to roll over on top of you, explode his load, and then leave you to sleep in a puddle. I mean, can't he sleep in the wet spot sometimes? The wet spot motivated me to initiate sex more often. If I could sidle over to his side and turn him on first, I could keep my side of the bed dry. Why does nobody talk about this? I honestly don't know. Maybe we're all desperate to uphold the image that sex is glamorous, over-the-top pleasurable, and movie-style tidy. The truth is that sex is slimy, messy, and leaves a wet spot. But it's worth it, right?

Is the wet spot pure, concentrated, medical-grade sperm? No. Not by a long shot. The sperm is mixed in with fluids created by the prostate gland, the seminal vesicles, and the bulbo-urethral glanads, which contain nutrients, emollients, enzymes, amino acids, and other elements that make sperm happy. Think of semen as the river that floats the canoe downstream.

If the wet spot bugs you, simply slide a towel under you during sex. When you're done, pull it out and, voilà, dry sheets.

I was a twenty-eight-year-old virgin when I first had sex with my boyfriend, but every time we have sex my vagina tears and my gynecologist tells me I have to wait a month before I can have sex again. Will I ever be able to have sex, or should I just buy myself a dozen cats and plan to grow old as a single, virgin spinster?

Yes, you'll be able to have sex, but it may take some time, determination, and patience. No two women have the same experience when it comes to losing their virginity. If you've been wearing super-plus tampons, riding horses, and straddling balance beams in gymnastics class, you might have no problem. On the flip side, if you're very tiny and trying to insert a junior-sized tampon sends you through the roof, you might face some challenges, especially if your partner is particularly well endowed.

My patient Adrienne faced a similar problem. When she met Hank, she thought he was the perfect guy. She and Hank shared a love of philosophy, a passion for all things spiritual, and a tendency to awaken at 5 A.M. Only one thing got in the way: Mr. Friendly. In fact, Hank's fraternity brothers nicknamed him Tripod. While men might believe that all women think bigger is better, this is not always the case. Every time Adrienne and Hank got freaky, her perineum tore in the same way some women tear from giving birth.

Adrienne would come to my office, and we'd spend the next four to six weeks working on healing her with sitz baths and ointments. Then, the next time they made love, boom. Another tear. Adrienne wound up asking the same question—would they ever be sexually compatible?

The answer is yes. Vaginas can be trained to accommodate guys like hung Hank. But it takes dedication. In Adrienne's case, I prescribed pelvic rest (no sex, vibrators, or tampons) to allow her fragile tissue to completely heal. Then, she committed to the following program (which can also be used for anyone wanting to lose her virginity as painlessly as possible).

HOW TO TEACH YOUR VAGINA TO LOOSEN UP

1. Purchase a good sexual lubricant like Astroglide or K-Y Jelly.

2. Begin to prepare weeks, even months in advance. The vagina is an elastic organ designed to fit a baby through it. Your virginal vagina may not like being stretched by an erect penis, but it's willing to negotiate. Begin by starting to dilate the vagina with the smallest thing you can insert into it comfortably—a pinkie finger, a junior-sized tampon, a Q-tip if need be. Find what just begins to stretch you but doesn't cause pain. Lube it up and gently insert it inside the vagina.

3. Notice the muscles that surround your vagina. Are they tense or relaxed? Does what you're inserting slide easily into the vagina, or does it hurt? Make a conscious effort to release the tension from those muscles.

4. Over time, try gradually increasing the size of what you insert into your vagina—a super-plus-sized tampon, two fingers, or a small dildo. Repeat the exercise with the larger objects until you find something that begins to stretch you and feel

uncomfortable. When you reach that point, hold the object still for ten minutes while reflecting on a memory that relaxes you—a vacation, a favorite resting spot, a childhood experience. Repeat this every day until it no longer hurts. (Don't forget the lube, which is key.) If you bleed a bit during some of these exercises, don't be concerned. If your hymen is not already broken, you may tear the fragile tissue around the vaginal opening, which may bleed a small amount.

5. When inserting the larger object no longer hurts, go nuts and go bigger. Keep increasing the size of what you place into the vagina until you can accommodate a dildo that mimics the size of your partner's penis.

6. Once you can keep the dildo in place for ten minutes without pain, try moving it in and out of you, mimicking the motions of intercourse. Go slowly, and make sure you have control of how the dildo moves. Feel free to play around with other types of sexual stimulation or sex toys during this process. Remember, sex is supposed to be fun! Feel free to invite your partner in for the party, if you feel comfortable. If not, make it a private affair until you're ready for the big night.

7. When you can do this without pain, you're ready for the main event. Encourage your partner to have fun with foreplay. The more aroused you feel, the less intercourse will hurt.

8. Spend a moment taking deep, relaxing breaths. Concentrate on relaxing the muscles around the vagina to make room for your lover to enter you.

Visualize your vagina as a flower gently opening to receive your partner.

9. When you're ready, lube up and try a woman-on-top position, which gives you more control over the depth of penetration and the speed of the thrusting. Take it slow. Listen to your body. Warn your partner that you may not be able to finish the job. If need be, you can ask him to stop and you and he will take care of each other in other ways.

10. Practice makes perfect. It gets better with time—I swear. You'll be making beautiful music in the sack before you know it.

So what happened to Adrienne and Hank? Lovemaking was gently and successfully consummated, without any tears. Then Hank broke up with Adrienne the next day. Why? He figured love just shouldn't be this hard, so he and his enormous wang moved on. I assured Adrienne not to regret it a bit. While she used to feel insecure about her narrow vagina, Tripod was proof that we can all stretch, grow, and become more accommodating in the name of love.

My husband says my vagina feels too loose and it doesn't stimulate him anymore, but I think he's just not getting hard enough. Can I do anything to tighten my vagina?

Vaginas evolve as our lives do. As virgins, we may be so tight that sticking a tampon in might hurt. But through the inevitable stretching that accompanies sexual intercourse, the vagina loosens up a bit. Childbirth, the ultimate exercise in vagina stretching, makes the vagina even looser. If you've delivered baby number four, you may be able to fit your whole fist in there without much resistance. Plus, as women age, the tissue changes and may lose some of its muscle tone.

What does this mean for all the penises out there? It can definitely have an effect, especially since many men face their own sexual challenges as they age. Your partner may have trouble achieving the kind of stiffy that came so easily when he was a young stud. So if you mix his limper erection with your looser vagina, you may wind up with, for lack of a better way to describe it, the equivalent of a hot dog penetrating a sweat sock.

So what can you do about it? First, determine where the problem really lies. Is it him? Maybe he needs a little help getting stimulated these days. I've never seen a man turn down a good blow job. If that still doesn't help him stiffen up, he may have more serious erectile dysfunction, which will require a trip to his doctor. Maybe a little Viagra is in order.

If his woody is raring to go, what's up down there for you? You might ask your gynecologist to check things out. Does your doctor need to use the humongous speculum we call "Big Bertha" in order to see your cervix? Do you have pelvic prolapse or a perineum that gapes open like a fish mouth? If this is the case, some Kegel exercises are on the Good Sex menu.

While you can't really tighten the vaginal tissue itself, you can tighten the pelvic floor muscles that surround the vagina. Just like lifting barbells or doing squats pumps up your biceps and quads, doing Kegel exercises increases the size and strength of the muscles that surround your vaginal opening. If Kegel exercises don't help, talk to your doctor. We have some other tricks up our sleeves to help loosey-goosey vaginas.

Do guys like it better if I make noise when we're having sex?

This question is *completely* personal. How a guy feels about the amount of noise you make in bed depends on a whole host of other factors. Does he have hang-ups left over from his childhood? Is he a devoutly religious believer in a faith that tries to suppress sexual impulses? Has he suffered sexual traumas in his past? What cultural beliefs does he harbor? Does he think all women should sound like the women in his porn videos? As you can see, there's no one answer to your question, so to get a better idea, I took a straw poll. Here's what some of my guy friends had to say:

> Nothing validates a man's lovemaking more than a woman who is enjoying it and is not afraid to express it in the form of sighs, screams, and vocal encouragement. Men die for this type of encouragement—the more and louder, the better for both partners. Sex is an expression of deep feelings of trust, passion, and love. Expressing it out loud, rather than hiding it inside, sends a wonderful message you *want* your loved one to hear.

> Hell yes! It lets me know I'm doing something right.

> Gotta say that, yes, it's gratifying (not to mention instructional, since each woman likes different stuff) to get aural feedback—to a point. If she's really flailing and waking the neighbors and sounding like Sinéad O'Connor riding a wild bull, then you begin to suspect that either she's faking it a little or she's having the kind of fun that means she's totally forgotten that there's a person attached to the other end of the cock.

I guess what I'm saying is that I like the noise, but I also like to know that she's still in the building with me.

Noise is good, but too much noise would make me uncomfortable.

Feedback is helpful; however, feedback that is too loud or sounds fake can be distracting.

The louder the better! I like a woman who knows how to scream!

While it's nice to hear she's having fun, excessively loud or inauthentic noise can be disconcerting and can definitely kill the mood.

Absolutely. It's really important for the guy. In so many ways, we need validation that what we're doing is working. Trust me, it's more fun for all of us if you give us some affirming noise.

Lou Paget says, "Women might think guys expect a lot of noise because they're comparing themselves to women in pornographic movies. But remember, those people are acting. It's scripted, they are voiced over, and there's someone giving them directions. In real life, what guys pay attention to is changes in your breathing and more subtle noises such as soft moans, which reflects how what they are doing is impacting your nerves and how aroused you are more than any porn star noises."

Why does my vagina make loud noises when my boyfriend and I are having sex?

Vaginal farts (some call them "queefs" or "varts") happen to almost all women at one time or another, especially during sex or other forms of exercise. Unlike gas expelled from the rectum, which contains fecal waste and may be stinky, vaginal flatulence is odorless and, unless a rare rectovaginal fistula exists, is completely unrelated to the rectum.

Why does this happen? When the vagina lengthens and the uterus shifts positions during sexual arousal, air may enter the vagina from the pumping action of intercourse or via oral sex. Air may get trapped where the vagina balloons out during this phase of arousal. Then, when you are getting into certain positions or when the walls of the vagina return to their unaroused shape, the air is expelled and, much to your embarrassment, the vaginal fart escapes.

What should you do about it? Just giggle, or say, "Excuse me." Sex is supposed to be fun. If you're letting your freak flag fly, there are likely to be other squeaks, squirts, and gurgles that erupt from a good romp in the sack. Don't let your embarrassment spoil the fun.

My boyfriend asked how many sexual partners I've had, and I'm embarrassed to tell him the truth. Am I a slut, or am I normal? What's the average number of sexual partners most women have had?

It's an awkward moment in any relationship when one of you asks, "So, how many sexual partners *have* you had?" The answer is pretty much guaranteed to make one or both of you feel like crap. If your partner has slept with dozens more people than you have, you may suddenly feel like just another notch on his bedpost. If you outnumber him, you may feel like a floozy. Either way, remember that what's done is done. You can't change the past. If you feel embarrassed or regretful about your past, you may be tempted to lie, but this serves no one. Lies only eat at you and come back to bite you in the butt.

Remember, if your boyfriend loves you for who you are, he has to realize that your past made you the kick-ass woman you are today. Tell the truth. If he really cares, he won't judge you for decisions you made before you were together. And remember, turnabout is fair play. If you confess your number and it turns out he's had more partners than you, no fair getting mad. It's all in the past and doesn't necessarily reflect who either of you are today.

Of course, another option is to simply say, "No good can come of this discussion—there is no right answer to that question, so let's avoid weirding each other out, and let the past stay where it belongs: in the past."

So how many sexual partners have most women had? Don't judge yourself based on what others have experienced. You're *you*. No one else is you. But to answer your question, the National Health and Nutrition Examination Survey, conducted by the National Center for Health Statistics, reports that the median number of lifetime sexual partners for women was four. Keep in mind that 25 percent of women reported only one lifetime sexual partner and 9 percent reported more than fifteen partners, so it all averages out. (In case you're curious, for men, the median number is seven, with 17 percent reporting only one lifetime sexual partner and 29 percent reporting fifteen or more partners.)[1]

Remember, don't kick yourself if you think you've had too many or too few. As long as you're being true to who you are right this minute, your number is just right for you.

*I want sex all the time, but my husband doesn't
seem the least bit interested. Am I the only woman
in the world who has to beg a guy for a little nookie?*

No, honey. You are definitely not the only woman who wants sex more than her partner. A recent survey conducted by *Woman's Day* magazine and AOL questioned 35,000 married women and found that 79 percent want sex more often, and only 19 percent call their sex life satisfying.[2] Michele Weiner-Davis, the author of *The Sex-Starved Marriage: Boosting Your Marriage Libido* and *The Sex-Starved Wife: What to Do When He's Lost Desire*, reports that 20 percent of all marriages have sex less than ten times per year.[3] Grim odds, eh?

One of my patients, Lily, was a virgin when she married her high school sweetheart, so she didn't have any other partners by which to evaluate her experience. After years of a nearly sexless marriage, Lily, who longed for a more intimate relationship with her husband, blamed herself and her self-esteem suffered. But there was absolutely nothing wrong with Lily. It wasn't until she separated from her husband and connected with a new lover that her sense of self blossomed.

Short of finding another lover, what can you do if your relationship is not as sexual as you wish? Dr. Michelle Gannon, San Francisco psychologist, couples therapist, and founder of Marriage Prep 101 workshops, offers some suggestions to improve your love life:

1. Let your partner know in a gentle, loving, non-complaining way that you would like your relationship to be more sexual.

2. Invite your partner to share thoughts on what might help him get in the mood.

3. Find out if there are certain times of the day that he feels more amorous. Make an effort to accommodate your sexual schedule to his.

4. Stress and fatigue can deplete sexual desire. Ask him to brainstorm how he might reduce the stress in his life.

5. Talk to him about masturbation. Some men may not have enough sexual juice to masturbate in the shower and still have sex with you.

6. Gently bring up the issue of porn. People who are addicted to porn may lose interest in real-live sex. If this is an issue, ask him to cut back or get help.

7. Remember that while half of the population need no external stimulation to feel sexy, the other half need to be kissed, touched sensually, and physically aroused before they feel turned on. This applies to men as well as women.

8. Keep in mind that decreased libido is a sensitive subject for most men, so be gentle with him. Many men are embarrassed if they are not interested in sex.

9. Read books together, attend an intimacy workshop, or see a counselor.

10. Last, but certainly not least, make sure the problem is not health related. If you can't tell, have him visit a doctor.

Lou Paget says, "The number one robbers of intimacy and sexual connection are fatigue, stress, and limited time. When men are stressed, especially regarding money, it completely wipes out libido. It may be that you have a naturally higher li- bido than your partner. Or maybe something (or someone) is in the way. If he was interested in sex before but things have changed, ask him why. If there's no one else and he's already done what he can to manage his time and stress levels, keep in mind that testosterone levels are lowest in the evening. Maybe it's time for morning sex."

Disparities between your partner's libido and yours can rob you of your self-esteem. Remember that, no matter what, you are beautiful, sexy, worthy, and lovable.

When my gynecologist asks me how many sexual partners I've had, how important is it that I tell her the truth? Do you docs really need to know our most intimate details?

Hmmm...I guess it's sort of important, but the exact number doesn't really matter to me. Here's what I really must know:

1. Have you ever had sex before?

2. Do you have sex regularly?

3. Are you at risk for sexually transmitted infections, pregnancy, and/or cervical cancer?

There are certain clinical decisions I would approach very differently if you've had three lifetime sexual partners as op- posed to thirty or three hundred. But if the real number is four instead of three, or thirty-five instead of thirty, to be honest, it doesn't really affect my decision-making. So if your partner is in the room and you never told him about lover number six, don't stress about it. Go ahead and lie. (I can't believe I just said that!)

Another question you may ask is whether we really need to know about that abortion you had that you've kept secret all these years and, frankly, never want to think about again. The answer for that is usually yes, especially if your visit relates to pregnancy or fertility. And that chlamydia you'd prefer to erase from your medical history is also important. So if you have secrets you wish to hide, visit your doctor alone or ask for some privacy during the examination so you can spill the beans without your boyfriend, Aunt Josephine, or your mother-in-law listening in.

When we ask these personal questions, it's not because we're trying to dig up the dirt and embarrass you. It's because the answers affect our treatment plans. Try to find a doctor you trust and remember, we're here to help you, not judge you.

I think I'm allergic to sex. Is that possible?

Yes, it's possible. While rare, it is possible to develop an allergy to the proteins in semen, a condition known as *human seminal plasma protein hypersensitivity*. Most cases result only in itching and swelling after sex, but some cases can be life-threatening. If you think you're allergic to semen, keep in mind that other allergens may come into play. Are you using condoms, lubricants, or spermicides? All of those can trigger allergic reactions, too.

How can you tell the difference? If every time you are exposed to semen in the absence of other possible allergens you develop vaginal itching and swelling—or possibly hives, wheezing, or anaphylaxis—you may have a semen allergy. Try using a condom. Do your symptoms magically disappear? If so, bingo.

Sometimes I bleed after sex when I'm not on my period. What does that mean?

If you're expecting your period any day now, sex may have stirred things up. Other than that, you'll probably require a trip to the gynecologist. While it's not uncommon, it's still not normal to bleed after sex in the absence of menstruation. Sometimes, harmless conditions can result in bleeding during sex, such as *ectropion* of the cervix (when the glands from inside the cervix are everted onto the outside of the cervix). Other times, bleeding after sex can result from a polyp, precancerous or cancerous changes of the cervix or uterus, vaginal warts, trauma, cervical infection (usually gonorrhea or chlamydia), vaginal atrophy caused by menopause or breast-feeding, or a whole host of other things you don't want. If you're bleeding after sex, especially if it happens more than once, alert your gynecologist. You gotta keep things healthy down there.

Lately, I started fooling around with a guy who has gotten around. He says he's clean, but I don't know if I trust him. Maybe he's just saying that to get in my pants? I obviously plan on using a condom when I sleep with him, but are there any visible signs of STDs I should look for before I agree to a booty call?

Honey, if you can't trust the guy, do you really want to sleep with him? I mean yeah, there are some things you can do to check him out, but it's not necessarily enough to protect you. Make sure you care enough about this guy that if you do get a sexually transmitted infection, it's worth it. Because the truth is, even if he's been tested and deemed "clean," you're still at risk.

Why? Because condoms don't protect you against all sexually transmitted infections and testing doesn't always test for everything. Most testing will not reveal whether a guy carries HPV, and often it will not tell whether he might be infectious for herpes, have pubic lice, or carry *molluscum contagiosum*. So what's a girl to do?

A FEW TIPS

1. Ask your partner to be honest about whether he has really been tested. Explain that you value the health of both of you and that if he cares about you, he needs to demonstrate this. Offer him a copy of your blood tests and ask for a copy of his if you can't trust him.

2. Give his junk a good once-over. Look for cauliflowery warts on his penis, scrotum, or around his anus that may represent HPV infection.

3. Inspect the tip of his penis for funky discharge (anything greenish or yellowish) that may indicate the presence of gonorrhea or chlamydia. Wetness in this region should always be clear.

4. Hunt for reddish ulcerations that might represent genital herpes.

5. Check for little round bumps caused by *molluscum contagiosum*.

6. Take a gander at his pubic hair to make sure there are no pubic lice or little white eggs.

Remember, a clean inspection doesn't mean you're good to go. Most sexually transmitted infections have absolutely no signs on a clinical exam. Which is why it all comes down to trust. At the end of the day, do you really want to hook up with someone you can't trust?

In my opinion, my partner and I almost never have sex. But when I told my friend how often we have sex, she said we were doing the deed way more than she and her partner were. On average, how often do couples have sex? What's normal?

Your questions remind me of a young woman I saw in my office. I asked her if she was sexually active, and she denied it. Then I diagnosed her with an unplanned pregnancy. Needless to say, I doubted that she was telling me the truth. When I pressed her again about whether she was sexually active, she said, "Well, not *active*. We only have sex three times a week."

It just goes to show you that everybody has a different idea about how much sex is enough. Take two people getting it on with the same frequency: once per week. One may be completely frustrated because she wishes she was doin' the bump daily. The other may be resenting the pressure from her partner and wishing she could scale it back to once a month. Truth is, we're all *so* different.

Since the word *normal* makes me cringe, I'll stick to citing statistics about what's average. According to the Kinsey Institute, couples have sex four times per week at fifteen to twenty years of age, three times per week at age thirty, twice per week at age forty, and less than once per week at age sixty.[4]

How much sex is enough? If you and your partner are both happy, it's enough. 'Nuff said. But if you or your partner is dissatisfied, it bears exploration, because those seeds of discontent will breed trouble ahead in your relationship.

Is there such a thing as too much sex?

Absolutely. If you are missing school or work, falling behind in your responsibilities, putting yourself at risk, or damaging your relationships because of sex, you may have a problem. Other behaviors associated with sexual addiction include repeatedly cheating on your partner, multiple or anonymous sexual partners, having one-night stands, consistent use of pornography, unsafe sex, frequent phone or computer sex (cybersex), prostitution or use of prostitutes, exhibitionism, obsessive dating through personal ads or the Internet, voyeurism, stalking, sexual harassment, and rape. If this sounds like you, please get help. There are treatment programs such as Sex Addicts Anonymous (SAA) that can help.

What do you think of porn for sexual excitement for women? If other women turn me on when I look at porn, am I a lesbian?

Many women are turned on by porn. And no, finding a woman sexy in a pornographic movie does not necessarily make you a lesbian. In fact, studies show that women's bodies respond sexually to other women. Even if women tell you they're not getting turned on by female porn, the physiology of their bodies betrays them. While straight men are not usually turned on by other men, straight women find other women sexy. Female sexuality is simply more fluid.

Until recently, I hadn't had much experience with porn. Sure, when my brother and I were kids, we would sneak upstairs to watch *The Sensuous Nurse* on HBO at 2 A.M. when my parents were sleeping. And my mother did give me that Playboy how-to video on my wedding night. But it wasn't a part of my sex life until recently.

After admitting that he thought a wee bit of porn might spice up our sex life, my husband searched for just the right erotic movie. After dallying around in the video store, looking over his shoulder for anyone he might recognize, he finally returned with the goods hidden in a bag.

So there we were, lying in bed, and these bad actors were sailing a pirate ship. They were mopping the deck, and the waves were crashing. Then all of a sudden, the pirate started ravishing the scullery maid, and everyone started moaning, and the TV screen filled with a close-up of something we...didn't exactly want to watch. We giggled, turned fire-engine red, and put in another movie. Maybe we just picked a dud.

The second movie pretended to be a crime drama. There was a detective in a trench coat, with his pretty girl Friday helping out. They were hot on a case, and they got a lead, but instead of chasing after the bad guy, Girl Friday whipped open her blouse, and the detective opened his trench coat, and, well... you can figure out the rest.

We were just about to write the whole thing off as a failed experiment when my husband confessed sheepishly, "Can I show you what *I* like to watch?" Turns out he has a subscription to what he considers a classy online site of downloadable video clips. I agreed to check it out. The first video we watched showed a beautiful woman lying on the beach in a gossamer gown, sunbathing. Then she got sweaty, took off her gown, and started rubbing lotion on her natural-looking small boobs, while soothing music played and waves crashed at her feet. I had to admit, it was pretty damn hot.

Am I a lesbian because I found the pretty girl in the porn video sexy? No. I like boys—always have. I'm telling you this story not to reveal too much about my love life (sorry, Mom!) but to remind you that we're all unique and we all respond to sexual arousal differently.

Used in the right context between two consensual adults, erotic movies can be a great way to spice up your sex life. But they're not for everyone. If it makes you feel uncomfortable, skip it. There are plenty of other ways to jazz up a lagging love life.

If I fantasize about having sex with a girlfriend, am I bisexual or a lesbian? Or just really close to my friends?

Feeling confused about your sexual identity can be unsettling, and if you've been raised to believe that homosexual thoughts are sinful, wrong, or inappropriate, fantasizing about having sex with a girlfriend may lead to feelings of shame, self-loathing, and fear of rejection. For any number of reasons, you may still be exploring your sexuality, and these thoughts are completely healthy and nothing to be ashamed about. The way I see it (and I'm not alone—famed sex researcher Alfred Kinsey agrees), sexual preference is not some black-and-white, yes/no answer. If you think about sexual preference on a numeric scale, with 1 representing someone who is completely heterosexual and 6 representing someone completely homosexual, many lie somewhere in between.

Klein Sexual Orientation Grid

Dr. Fritz Klein wished to expand upon Kinsey's scale and developed the Klein Sexual Orientation Grid.[5] To fill out this grid, ask yourself the questions and fill out each box in the grid with a value between 1 and 7. For A through E, the possible answers are 1 = Other sex only, 2 = Other sex mostly, 3 = Other sex somewhat more, 4 = Both sexes, 5 = Same sex somewhat more, 6 = Same sex mostly, and 7 = Same sex only. For variables F and G these range from 1 = Heterosexual only to 7 = Homosexual only.

	PAST (ENTIRE LIFE UNTIL NOW)	PRESENT (LAST 12 MONTHS)	IDEAL (WHAT WOULD YOU LIKE?)
A. Sexual Attraction: To whom are you sexually attracted?			
B. Sexual Behavior: To whom have you actually had sex?			
C. Sexual Fantasies: About whom are your sexual fantasies?			
D. Emotional Preference: Who do you feel more drawn to or close to emotionally?			
E. Social Preference: Which gender do you socialize with?			
F. Lifestyle Preference: In which community do you like to spend your time? In which do you feel most comfortable?			
G. Self-Identification: How do you label or identify yourself?			

Each of the twenty-one boxes should contain a value from 1 to 7, categorizing your answer to each question. How you answer these questions may help you assess whether or not you are bisexual. If you're confused about how to interpret the results, talk to a therapist.

Most important, how do you identify yourself? It's not whether other women turn you on, who you fantasize about, or even how you behave. It's how you feel inside. As you grow older, your sexuality may evolve. Don't be afraid of what you feel.

If you know deep down that you're bisexual or a lesbian, you may be scared to acknowledge these feelings because you fear isolation, loss of friends and family, and other societal conflicts, which may be very real. If you feel confused or ashamed, or you're in the process of coming out, consider talking to a counselor who can help you work through your feelings.

I've heard there are health benefits to having sex and orgasms. Is this true?

Yay for us! That would be a resounding yes. In addition to the giddy euphoric effects that make our toes curl, sex and orgasm (including masturbation) seem to have other health benefits. Beverly Whipple, Ph.D., R.N., famed sex researcher and professor emerita from Rutgers University, lists the following evidence-based benefits of sexual expression. Engaging in acts of sexual expression may:

1. Help you live longer[6]

2. Lower your risk of heart disease and stroke if you have sex twice a week or more often[7]

3. Reduce your risk of breast cancer[8]

4. Bolster your immune system[9]

5. Help you sleep[10]

6. Make you appear more youthful[11]

7. Improve your fitness[12]

8. Help protect against endometriosis[13]

9. Enhance fertility[14]

10. Regulate menstrual cycles[15]

11. Relieve menstrual cramps[16]

12. Help carry a pregnancy to full term[17]

13. Relieve chronic pain[18]

14. Help reduce migraine headache pain in some individuals[19]

15. Improve quality of life[20]

16. Reduce the risk of depression[21]

17. Lower stress levels[22]

18. Improve self-esteem[23]

19. Improve intimacy with your partner[24]

20. Help you grow spiritually[25]

The evidence is mounting. Orgasm isn't just good—it's good *for* you.

Sometimes I'm very aroused, but my vagina is still bone dry. Why is that?

I hear you. You're in the middle of a hot, steamy sex scene and suddenly it's time to go at it, but instead of feeling slippery and ready for sex, your ya ya feels like the Sahara Desert. What's up down there?

You might assume that your vagina will automatically lube right up when you get excited, but that's not always the case. Vaginal lubrication during sex begins during the excitement phase, which is the first stage of the sexual response cycle described by famed sex researchers Masters and Johnson. As you become excited, blood rushes to the vaginal walls and begins the process of lubing you up. As you reach the plateau phase, glands produce further lubrication that makes you wetter still. While much of this physiological process is influenced by your mind, your hormones play a role as well. Your mind may be raring to go, but if your hormones aren't on board, you might wind up feeling aroused while your vagina is telling a different story.

As we age, levels of estrogen may drop, resulting in vaginal dryness, even when you're turned on. If vaginal dryness is getting in the way of your sex life, try using olive oil or an artificial lubricant like Astroglide to simulate the natural lubrication your body isn't making. Or talk to your gynecologist about whether estrogen, either systemically or vaginally, may improve your sex life.

What do women like in bed? My boyfriend is asking and I'm too embarrassed to talk about it. Can you help?

The question isn't what I think women like in bed. What he's really asking is what *you* like in bed. Count your blessings. This guy cares what you think! Throw out your inhibitions and tell him the truth. Here are few generalizations you might use as guidelines, but don't take my word for it. Write your own sexual manifesto. The truth will set you free.

A CHEAT SHEET FOR GUYS

1. Every woman is different. If your super-duper signature technique had your last girlfriend curling her toes and bellowing out to Mother Mary, good for you. But don't expect the same thing to work on your new lover. Our bodies—and needs—vary drastically. One size does not fit all.

2. A woman's body is like an old beater car in subzero weather. It takes a while to warm us up. Don't expect a warm welcome if you zip straight to the vagina without a little foreplay. Our bodies sometimes need a little coaxing. Too often, we live completely in our heads. Our minds are spinning with thoughts about work, the kids, and tomorrow's to-do list. If you help bring us into our bodies by arousing different erogenous zones, like the ears, the lips, the breasts, the inner thigh, the belly button, even the toes, you help remind us that our bodies can offer pleasure if we only inhabit them.

3. Love us and earn our trust. For most women, sex and love get all tangled. Not to say there aren't some out there who enjoy just getting it on for the sake of sex. But most of us see sex as an expression of love, and if we don't feel nurtured, we may not get all hot and bothered when you want to shake the sheets. Treat us tenderly and pleasure will likely follow.

4. Set the mood in the bedroom. Surprise us with candles, mood music, and a flower on the pillow. Whisper sweet nothings. Don't serve up silly platitudes, but say what you feel. When we cover our bellies with our hands and try to turn off the light, tell us we're beautiful, just the way we are. Share how much you care. Romance gets us in the mood and helps us relax.

5. Know a woman's anatomy. Need help? Take the Pretty Pink Pussy Tour (pages 60–61).

6. Think sensually, not sexually. Immerse yourself in the sensory experience and find your own timing together.

7. Give us permission to offer feedback, and don't take it personally. If we don't respond to something you're doing, it doesn't reflect on your skill as a lover. It just doesn't work for our unique anatomy and physiology. If you act dejected every time we offer you feedback, we're likely to stop trying to help you please us. Accept constructive criticism lovingly.

8. *Never ever* compare us to another woman. We don't care what the hell Jill or Sally or Maryanne liked in bed. If you think about other women when you're making love to us, please—for the love of God—keep your thoughts to yourself.

9. Fine-tune your radar. Even if you invite us to offer feedback, we may not feel comfortable talking about sex. Many of us have been so conditioned to consider sex taboo that we clam up when the subject arises.

Learn to read our subtle signals and over time you will discover what pleases us. Little grunts, moans, and heavy breathing usually signal *yes,* and while silence may simply signal shyness, it may also mean that what you're doing isn't working. Pay attention to body language, too. When we move toward you, it's a good sign, and if we adjust our body to a different angle, we might be trying to show you where we want you to be.

10. Be gentle and go slow. There's no race to the finish line here. Remember how sensitive girl parts are. Don't mash on us (unless we ask you to!). Start slow, then gently pick up the pace as you go. Don't start banging us around like you're trying to get to home base before we've even gotten up to bat. You may get sprung in ten seconds flat, but chances are, we're still thinking about how little Johnny's teacher thinks he needs a reading tutor, or whether we're prepared for that big presentation at work tomorrow. Be patient with us and our monkey minds.

11. Pull out the Kama Sutra. No need to focus all your energy on making us orgasm during intercourse, but why not try? Check out some books about sexual positions and have fun experimenting. You never know what might hit the spot for your lover. Be creative.

12. Don't take it personally if we don't orgasm during intercourse. Some lucky women get off from the mere thought of intercourse, but the majority of women do not orgasm through intercourse alone. If you expend

so much energy trying to make us come during intercourse, you may miss the rich opportunity to satisfy us in other ways. Sure, try your darnedest to please us. But don't pressure us. Many women will not orgasm during intercourse, even with the most skilled partner.

13. Help us out. If your lover prefers to orgasm during intercourse, start with oral sex to help sensitize her delicate organs. Encourage her to explore positions that stimulate her clitoris, such as the woman-on-top position. Use your hands to touch her while you're having intercourse, or invite her to touch herself. She knows best what feels good, and if you tell her how much it turns you on to see her touch herself she may feel more comfortable augmenting her own pleasure.

14. Most women love oral sex. To a woman, it just doesn't get much better than this. Soft, wet tongue meets delicate pink pearl. Can you hear us purr? We love it even more if we think you do, too. Start gently. Explore the inner thighs, the labia, the opening to the vagina. When body language indicates that we're ready, lick, suck, and swirl the clitoris in circles, mixed with up-and-down motions. Use your hands to explore the rest.

15. Just because you're done doesn't mean we are. If you come before we do, no stress. Just finish the job and help us feel as good as you do.

16. Invite sex toys into the bedroom. The sex toys are your friends, not your competition. Let them stimulate both of you, and encourage her to explore.

17.   Get tantric or explore Taoist sexuality. In addition to deepening your connection to your partner, elevating your lovemaking to a spiritual plane may contribute to your spiritual growth together. Check out *The Multi-Orgasmic Couple: Sexual Secrets Every Couple Should Know,* by Mantak Chia, Maneewan Chia, Douglas Abrams, and Rachel Carlton Abrams, M.D., for exercises in Taoist sexuality. Or try *Urban Tantra: Sacred Sex for the Twenty-first Century,* by Barbara Carrellas. If you're interested in tantric sexuality but short on time, read *Tantric Sex for Busy Couples: How to Deepen Your Passion in Just Ten Minutes a Day,* by Diana and Richard Daffner.

18.   Remember that sex is about making love. Don't get so focused on technique that you forget to connect. Look deeply into our eyes. Caress us lovingly. Tell us how you feel. Hug us. Love us.

19.   Cuddle when it's over. Please don't jump up and go watch the game. We make ourselves vulnerable, put ourselves out there, and want to know you're still with us when it's over. Snuggle in and stick around awhile.

CHAPTER 5

Orgasm

YOU PROBABLY REMEMBER THE INFAMOUS scene from *When Harry Met Sally*, when Harry and Sally are arguing about women who fake orgasm. Sitting in a crowded restaurant, Harry swears he would be able to tell if a woman faked orgasm and Sally insists that he wouldn't. She then proceeds to twist and moan, sighing with pleasure, finally climaxing with a resounding, *"Yes! Yes! Yes!"* while everyone in the restaurant watches. Director Rob Reiner's mother delivers the best line in the movie, saying to the waitress, "I'll have what she's having."

If only real orgasms were that easy. The truth is, orgasm is only one small piece of a very large life puzzle. But ah...what a piece!

At a workshop, I was sitting in a circle with a group of women and I invited them to share their experiences about the first time they discovered orgasm. One woman blushed and confessed that she was sunbathing on a lawn chair in her backyard when a fly started crawling on her inner thigh. The delicate touch of this fly on her leg gave her tingles, and before she knew it she was lying there alone in her bathing suit, reveling in the rush of orgasmic shudders. Another was hanging by her arms off the edge of the swimming pool, positioned just so in front of one of the jets spouting water in exactly the right spot. She became a fish that summer and mourned the first frost. One woman was merely fantasizing about having Brad Pitt when the tingling sensation that started at her crotch and spread through her body like an electric shock sent her through the roof. Another forty-six-year-old woman had no idea what an orgasm even felt like.

I may not be like some women, who can orgasm from a fly or the mere thought of Brad Pitt, but I can usually find my O with the tried-and-true methods of clitoral stimulation. Sadly, though, I've never been one of those 30 percent of women who can get off on intercourse alone. After years of watching Sally and other romantic heroines explode in rapturous bliss, I went through a phase where I was convinced I would be able to orgasm during intercourse if only I approached sex the way I approached medical school—with pure hard work. I didn't confess my secret obsession to my partner. After all, if he knew my goal, he would also know when I failed. Instead, I blundered off on a stealth mission to find my elusive orgasm during intercourse.

I read *Cosmo* articles and books. I initiated new positions during sex. I rocked and tilted and visualized. But alas, my orgasm stayed in hiding. After a while, disappointed at my F in lovemaking, I gave up. I figured sex just shouldn't be that much work. And my partner was happy to help me find other ways to get off.

Years later, I was with a guy who really didn't deserve me.

He was cute and sexy but all-around bad news. We had been rid-ing the pony together for a few months, and frankly, it was not all that. Then one night, after a few too many glasses of wine at a dive bar where croaky-voiced smokers sang the blues, it came. The ever-elusive intercourse orgasm arrived, a giddy and unexpected ecstasy. That moment required no special position, no toys, no elaborate tantric rituals, no magic technique. It just happened, and then, like a wisp of a cloud in a sunset sky, it was gone.

That was eleven years ago, and it has never happened since. I've replayed the scene in my mind and can't think of a single thing we did differently. I'd like to say it was this profoundly meaningful experience that changed my life. But frankly, the guy dumped me on my birthday a short while later, so I still feel a little sour grapes about the whole thing. To be honest, I wish I could take it back. But alas, just as orgasms are fleeting, these moments in time come—and then they go—and you can't undo the past.

I learned one good lesson from the whole thing: You can't *make* an orgasm happen. It's not something you *do*. Like much of life, it's something you must surrender into. You don't effort your way to orgasm; you simply let go.

Now, I no longer care how my orgasm arrives. Oral sex, manual stimulation, sex toys—it all works for me, so why stress out about something that's supposed to be fun? I try to release expectation, allow my body to feel what it feels, and enjoy the fleeting fireworks however and whenever they come.

What percentage of women have orgasms?

In the United States, 40 percent of women express concern about their sex life.[1] And those are just the ones who admit it! Among those with sexual concerns, orgasm ranks right up there. Eighteen to forty-one percent of women in a worldwide study reported the inability to experience orgasm, so if you do the math, 59 to 82 percent of women report the ability to experience orgasm.[2]

What does an orgasm feel like?

I'm not sure how to articulate in words something so visceral, but I'll try. It's like a momentary suspension in time, which begins as a slow buildup of tingly, pleasurable feelings that increase your sensitivity and culminate in a sort of sneeze, which begins in the clitoris and spreads throughout the body, resulting in rhythmic pulsing contractions in the vagina that leave you with a general sense of contentment and well-being. It's a giddy, exhilarating feeling, like champagne bubbles fizzing up and popping within you. But none of those words quite do it justice.

I asked my girlfriends to help me on this one, and here's how they described orgasm:

1. *"Like a roller-coaster ride. It takes a while going up, and it's a fun ride down."*

2. *"Like an internal explosion, sort of like a volcano erupting."*

3. *"You become warm and fuzzy all over, and then, for a brief moment, you become one with the Universe."*

4. *"A good orgasm feels like an out-of-body experience, like a 'little death' but in a good way."*

5. *"The ultimate tension release."*

6. *"Warmth, glow, adrenaline rush, calm, and bliss, all at once."*

7. *"You are transported into a secret world of wonder, ecstasy, and fantasy. At that moment, the mind stops, there are no questions, and there are no answers. You feel the ultimate availability of unharnessed life force energy exploding through your body in wild abandon."*

8. *"Your whole body floods with heat all from one starting point."*

9. *"Your entire being relaxes into receiving and feeling every moment of pleasure and love for yourself. You plug in."*

10. *"Animal longing transforms into pulsating spirit of pleasure. You feel warmth, certainty, the ancient call, with yummy-yummy rippling through your sex, belly, heart, and brain. Your yoni flows out and in, like a veil dancer, giving, then giving a bit more. You're on the edge. You're rosy pink alive. You breathe. You pant. You cry out. Then finally, you come home."*

One friend pretty much summed it up, saying, "Writing about orgasm is like dancing about mathematics." Amen to that.

I can masturbate just fine, but I can't seem to have orgasms of any sort when I'm with someone else. What's wrong with me?

Nothing's wrong with you, sweetie. But there may be things you can do to feel more right. While it may not feel like it, it's very good news that you're still able to orgasm through masturbation. If you can orgasm with masturbation but not with a partner, at least we know your parts are all in working order. Some women can't orgasm either way, which can signal physical problems, such as nerve damage or circulatory problems. But if you can get your juices flowing by yourself, we know that's not the case.

Most women who can't orgasm with a partner are able to orgasm alone, which signals a psychological barrier of some sort. Things like anxiety, depression, stress, embarrassment, guilt, alcohol, drugs, and fear of pregnancy or sexually transmitted infections may erect roadblocks to orgasmic bliss. Other issues, such as feeling disconnected from your partner, lacking trust, the presence of unresolved relationship conflicts, or a history of sexual abuse, can also get in the way.

In addition to these common barriers to orgasm, many of us also suffer from religious or cultural hang-ups we learned as kids. If you still hear Great-Aunt Gertrude's voice saying, "Only bad girls enjoy sex," or, "Sex goes against God's will, and God is always watching," it's no wonder you freeze.

If you feel safe and comfortable, try masturbating in the presence of your partner. If you're in control and your partner is supportive, you may be able to bring yourself to orgasm, which is the first step toward experiencing orgasm in other ways. Most of the time, you can learn to overcome these challenges, unfolding the flower within you. Talk to a good gynecologist or—even better—a sex therapist. We can help.

I've never had an orgasm in my life. What's wrong with me?

You are one of the 10 to 15 percent of women who have a condition we call "primary anorgasmia." But don't despair. The odds are good that you're capable of having an orgasm, if you care enough to commit to it. Maybe you're taking medications that get in the way, or maybe health issues are a problem. Drugs and alcohol can also numb you and lead to difficulty achieving orgasm. Or maybe you're not getting enough attention right where you need it, if you know what I mean.

More commonly, problems with achieving orgasm result from psychological causes. Remember that you can't control an orgasm. You must release into it. For some women, this may require a very special set of circumstances.

Take Clarice, for example. Clarice was forty-nine when she confessed, for the first time in her life, that she had never experienced an orgasm. When I probed further, she admitted that she and her husband of twenty-one years were frequently intimate—once or twice every week. She had no problems feeling aroused, became lubricated when she was stimulated, and enjoyed intercourse with her husband for the intimacy it bred between the two of them. But when it came right down to it, she didn't know what all the fuss was about.

When I asked if she was able to achieve orgasm through oral sex or masturbation, she turned bright red and said, "We don't do those things." Manual clitoral stimulation, sex toys? She shook her head again and said, "We're kind of old-fashioned."

When I asked if it bothered her not to have orgasms, she shrugged and said that she'd never really questioned it. But after thinking about it for a few minutes, she admitted that an orgasm would be nice. I explained that the most important thing we needed to figure out was whether her problem was physiological, psychological, or technical. In other words, was something wrong with her body that prevented her orgasm? Was she experiencing performance anxiety or recalling some past abuse? Or, even more simply, were she and her husband just not stimulating her where she needed to be stimulated? The best way to sort this out is to learn to masturbate.

Clarice blushed every time I mentioned the word. Because she had never masturbated and found it very embarrassing to discuss, I gave her a handout I had written, explaining, in detail, with hand-drawn body parts, diagrams, and arrows, how I wanted her to touch herself. First, I prescribed a hot bath, candles, and stimulation of her nipples. When she felt comfortable, I told her to turn down the lights in her bedroom, lock the door, and play her favorite music. I suggested that, giving herself plenty of time, she experiment with touching herself, finding the places she enjoyed touch and giving herself permission to indulge these feelings. When she was ready, I suggested that she try rubbing her clitoris with her fingers to see if stimulating her clitoris directly resulted in orgasm.

Although she resisted my suggestions at first, she ultimately acquiesced. Finally, through masturbation, Clarice discovered that—ooh la la!—she was able to achieve orgasm, which ruled out any sort of physiological reason for her problem. I suggested that she and her husband could see a sex therapist, if they were interested in exploring this issue further. Clarice didn't feel comfortable suggesting this, but over time she expressed her desires to her husband, who was shocked and disappointed that she'd never experienced orgasm before. He always assumed she had orgasms but was too modest to express herself vocally. When she admitted that she'd never had an orgasm in twenty-one years of sex, her husband studied new techniques, and I'm happy to say, Clarice has finally found her voice—and her O.

I can have orgasms, but never during intercourse.
Why is that? Am I the only woman who isn't
hanging from the chandeliers during sex?

If you survey women, only 30 to 35 percent will orgasm during intercourse, and among these, most need to work at it.[3] Many of us are looking for the "Look, Ma! No hands" orgasm, but few women are that lucky. While a few report the ability to orgasm just by thinking sexual thoughts or having their breasts touched, most women, even most of those able to achieve orgasm during intercourse, need direct clitoral stimulation to push them over the edge.

Some sexperts believe that all orgasms result, in some way, from clitoral stimulation, whether it's from masturbation, oral sex, a sex toy, or vaginal intercourse. Most women who are able to orgasm from intercourse alone have learned how to put just the right pressure on the clitoris. The truth is that the majority of sexual positions that involve vaginal penetration fail to stimulate the clitoris at all. Most women need something more—stimulation of the shaft or glans of the clitoris or friction or rubbing against the clitoral hood. When you stimulate this area, the pudendal nerve transmits sexy sensations to the spinal cord, which passes them to the brain, where the impulses are interpreted as pleasure. The pleasure signals get sent back down the spinal cord, via the pudendal nerve, back to the where the sensation started.

While some swear by the G-spot and vaginal orgasms, experts still argue. The way I see it, who really cares where your orgasms come from, as long as you're having them, one way or another? Don't get so caught up in *how* you experience orgasm that you forget to have fun.

Does the G-spot really exist?

According to the teacher in my Gross Anatomy lab, the answer was no. As we were dissecting the vagina, someone asked, "So where's the G-spot, Doc?" My teacher, in his thick Eastern European accent, said, "Zere is no G-spot in ze human female." Okay, good to know.

The rest of my medical training pretty much agreed with Professor Von Buzzkill. An expert in the field even told me that every part of the vagina has been examined under the microscope and there is nothing on the anterior wall of the vagina that looks any different than the rest of the vagina. Therefore, the G-spot does not exist. Period.

However, as is the case with much I learned in medical school, my patients tell me otherwise. Over the years, thousands of patients have sworn that there is a place felt through the anterior wall of the vagina that hits the oh-oh-oh spot—or, rather, *is* the spot. I believe in many things I cannot see, so I tend to believe my patients.

Hunting for data to validate their experience, I came across Dr. Beverly Whipple, who famously named the G-spot after German OB/GYN doctor Ernst Gräfenberg, who described a zone of erogenous feeling on the anterior wall of the vaginal canal. (A friend of hers suggested she name it the "Whipple Tickle," but out of respect for Whipples everywhere, she vetoed this idea.) According to Dr. Whipple, the G-spot definitely exists. When I asked her why some in the medical community vehemently deny its existence, she seemed baffled. She said, "I don't know. I guess, because they can't see it under a microscope, they think it doesn't exist. But my career has been about validating what real women experience. And some—but not all—definitely experience pleasurable feelings when you stimulate the G-spot area."

Her belief runs so deep that she went on to conduct hundreds of studies aimed at validating the sexual experiences women relate. For one study in 1981, four hundred female volunteers were examined. According to Dr. Whipple, a spot that empirically swells with stimulation was found in each of these women, although she admits that not all women appear to be sensitive to this type of stimulation.[4]

So what is the G-spot? Dr. Whipple isn't sure. As Dr. Von Buzzkill said, no specific anatomic differences can be detected in this area. But she suspects a cluster of blood vessels, nerves, glands (including the "female prostate gland"), and part of of the clitoris may all merge to create a sensitive area that hits the spot. She believes the female experience more than the microscope, and I tend to agree with her.

Where is the G-spot?

Those who study the G-spot say it lies two to three inches inside the vagina, on the anterior wall (the one closest to your belly button, not your bum), just under the urethra. They describe this area as having a different, ridgy texture from the rest of the vaginal landscape. Women tell me that the sensations they experience from stimulation of the G-spot differ vastly from those they feel with clitoral stimulation. While the clitoris is much more sensitive and more easily aroused, the G-spot requires deeper stimulation, which supposedly results in deeper orgasms. Is this true? Too many women say it is, and I'm more inclined to believe them than Professor Von Buzzkill.

Why can't I find my G-spot?

If you've read the manuals, tried all the techniques, and can't seem to locate your G-spot, I'm with you, girlfriend. I am one of the *many* women who cannot personally find their spot. Frankly, the clitoris works just fine for me, thank you very much, but I'm all about sexual exploration. Own your sexuality—investigate ways to achieve multiple orgasms, work your way through the Kama Sutra, and by all means, find that elusive G-spot. Happy hunting!

It's possible you'll find your G-spot and discover that stimulating this area gets your motor running, but you can't quite achieve orgasm. No worries. The G-spot may add to your sexual pleasure, but you may still need clitoral stimulation to push you over the orgasmic edge. Think of the clitoris and the G-spot as the whipped cream *and* the cherry on top. No need to choose.

If you can't find your G-spot, don't feel bad. You're not alone. Dr. Beverly Whipple says, "If G-spot stimulation feels good, then women should enjoy it, but they should not feel compelled to find the G-spot."

While some women say the G-spot allows them to experience vaginal orgasms from intercourse alone (usually from positions that angle the penis so that it hits the anterior wall of the vagina, such as doggie-style), this is rare. While some women swear they can orgasm through intercourse alone, it's usually because they've mastered the art of positioning themselves and their partners so that the clitoris is stimulated directly or indirectly.

Don't forget that the ultimate goal of sex is intimacy. You might get so caught up in G-spot hunting that you forget to have fun and express the loving connection sex is all about. If you're feeling sexually satisfied, don't let yourself or your partner stress about finding some proverbial holy grail so much that it keeps you from enjoying the wine.

Sometimes, when I'm really turned on, I soak the bed with fluid when I orgasm. Why do I squirt when I orgasm? Is it pee?

Yes, it may be pee. But it may be something else. The elusive and controversial "female ejaculation" elicits loads of speculation. The truth is, nobody seems to know for sure what's up down there when this happens. What we do know is that you're not alone in experiencing it.

Belle was having sex with some dude she barely knew when her body decided to pull out its newest parlor trick. In the middle of a doggie-style romp in the hay, she found herself spraying fluid all over the sheets right as a rip-roaring orgasm that left her in tears shook her to the core. When the orgasm passed, she found her partner staring at her, openmouthed. He said, "What the hell just happened?" She wanted to crawl under the bed and disappear.

She showed up in my office, repeating his question: "What the hell just happened?"

I had to shrug my shoulders and admit that nobody knows for sure what really happens when a woman gushes fluid during orgasm. Some insist these women, in the surrender of ecstasy, are merely incontinent, leaking urine from the urethra during the release of orgasm. Others swear the fluid comes from the paraurethral glands, which are located in the lower part of the urethra. Perhaps both theories are true. Those who squirt report the expulsion of anywhere from a teaspoon to a cup or two of fluid.

You'd think we'd know the answer by now, but sex researchers can't seem to agree. Biochemical analyses of the fluid show variable results. When studied, some "ejaculated" fluids look like urine, while some don't.[5] So is female ejaculation fact or fiction?

Dr. Beverly Whipple believes it is fact. Her studies of "female ejaculate" show that the fluid released contains glucose, fructose, prostatic acid phosphatase, and PSA, substances not normally present in urine. What confuses matters is that some women are, indeed, incontinent during orgasm, but this is distinctly different from female ejaculation. The fluid resembles fat-free milk, tastes sweet, and rarely exceeds a teaspoon in volume. If you're soaking the bed with cups of fluid during orgasm, it's probably urine.

According to Mary Roach in *Bonk: The Curious Coupling of Science and Sex,* one woman devised a home experiment to test for herself. After swallowing pills that dye urine blue, she inspected her wet spots, which were either colorless or faint blue. Of six laboratory evaluations of ejaculated fluid, two concluded that the fluid was urine. Four found significant differences between the fluid and urine.[6] And so the debate continues.

Regardless, I say if you're a squirter, embrace it. If guys can squirt fluid when they come and not feel self-conscious about it, so can you. And if you don't ejaculate, don't sweat it. You're normal, too.

Why can some women have multiple orgasms when men can only have one?

After ejaculation, most men experience a refractory period, during which they are physiologically incapable of achieving an erection or ejaculating again, no matter how much they are stimulated. (Although Dr. Beverly Whipple studied one man who could ejaculate repetitively, with no apparent refractory period, this is the exception, rather than the rule.[7]) How long the refractory period lasts varies widely between men and tends to be age dependent. Young men have brief refractory periods, while it may take much longer for an older man to achieve an erection after ejaculation. While there's no way for a man to avoid the refractory period after ejaculation, multiple orgasms are possible for men if they can delay ejaculation, a technique taught by Mantak Chia and Douglas Abrams in their book *The Multi-Orgasmic Man: Sexual Secrets Every Man Should Know.*

On the other hand, women do not experience this physiological refractory period. However, the clitoris may be so hypersensitive after orgasm that multiple orgasms are difficult to achieve. For those women able to tolerate continued stimulation after one orgasm, multiple orgasms are possible.

My girlfriend claims she can have eight orgasms in a row. We call her Octo-O! Is she lying?

Is your girlfriend trustworthy? While I can't say for sure whether Octo-O is pulling your leg, it's certainly possible for a women to have eight consecutive orgasms, especially if she has been practicing sexual techniques designed to allow for such bliss.

How do you define multiple orgasms? Good question. At its most basic definition, it means more than one orgasm during one lovemaking session. But within that definition lies great variety. According to Susan Crain Bakos, author of *The Sex Bible: The Complete Guide to Sexual Love,* women can experience multiple orgasms in multiple ways.

1. **COMPOUND SINGLES.** Each orgasm is distinct and separated by a partial return to the resolution phase. If orgasms are pearls, imagine a pearl necklace, with a segment of gold chain dropping down between each pearl, representing a drop-off in sensation.

2. **SEQUENTIAL MULTIPLES.** These orgasms may be two to ten minutes apart, with minimal reduction in sensation in between. In this scenario, the pearls are separated by a small segment of taut chain.

3. **SERIAL MULTIPLE ORGASMS.** Numerous orgasms are separated by mere seconds or minutes at most, with no diminishment of arousal. Some experience these types of multiple orgasm as a lone long orgasm with spasms of varying intensity. In this scenario, the pearls are right next to each other, with little to no separation between them.

Ultimately, each woman's orgasm is completely unique, an "orgasmic fingerprint" of sorts. So chances are that's just Octo-O's particular orgasmic style.

I masturbate a lot and now it seems like it's harder for me to have an orgasm when I'm with my partner. Am I making my clitoris insensitive by rubbing it so much? Do you see any problems with women who are too reliant on their vibrators to have orgasms?

You're not damaging your clitoris by rubbing it. But you may be increasing your pleasure threshold by masturbating a lot. If you never knew how pleasurable masturbating could be, your body might be satisfied with less stimulation. By training your body to expect a certain level of ecstasy you may find it harder to achieve orgasm during partnered lovemaking.

Keep in mind that you know exactly what works for you, whereas your partner may be blindly guessing what you like. Plus, I have yet to meet a human who can compete with a vibrator. The truth of the matter is that simple thrusting action from your partner just isn't going to do it for most women, especially if your body expects more. But that doesn't mean you can't bring your vibrator out to play during partnered lovemaking. Let your partner know that it's not a competition. You, your partner, and your vibrator can all have some rockin' good fun in the bedroom together.

Antidepressants make me unable to have orgasms. And I've noticed that my depression worsens when I'm not sexually active. Is it possible to be happy and have orgasms at the same time?

Yes, it's absolutely possible to be happy and have orgasms. Antidepressants are famous for being buzzkills. Decreased libido and difficulty achieving orgasm are well-known side effects of many antidepressants. If you're hoping to be happy and have orgasms, take a look at the reason you're on antidepressants. I have nothing against antidepressants. In fact, I've seen them save many lives. But too often, I see patients being treated for natural occurrences—like menopause or grief—using antidepressants when the real underlying issues are not being addressed.

In these cases, hormone balancing or psychotherapy can help you work through the issues that may be causing your depression. So if antidepressants are robbing you of your sexual life, reconsider whether they're really necessary.

As Julia Ross, M.A., explains in *The Diet Cure: The 8-Step Program to Rebalance Body Chemistry and End Food Cravings, Weight Problems, and Mood Swings—Now* and *The Mood Cure: The 4-Step Program to Take Charge of Your Emotions—Today,* simple nutrition strategies can often raise serotonin levels naturally while avoiding drug side effects.

I have found in my holistic women's health practice that strategies aimed at improving nutrition, encouraging exercise, optimizing natural serotonin production, reducing stress, and looking inward to heal underlying wounds may eliminate the need for antidepressants. But not always. Sometimes depression is simply biochemical. If antidepressants improve your quality of life immeasurably and you're one of those people for whom they're truly lifesaving, talk to your mental health provider about switching antidepressants or experimenting with your dosage.

While some data suggests that Viagra may help counteract the sexual side effects of antidepressants in women, Viagra has not been FDA-approved for use in women. Gingko biloba has also been reported to counteract the sexual side effects of antidepressants in both men and women.[8] How else can you get off without getting off your antidepressants? Experiment with ways to spice up your sex life (see pages 88–89). If you must sacrifice orgasms to stay sane, don't despair. As the authors of *The Orgasm Answer Guide* say, "Orgasm is an *experience,* not a goal.... Many people have satisfying sexual experiences without orgasm."[9]

The resilience of the human spirit continues to leave me in

awe. We can overcome any loss and still wind up happy. Sexuality for women is much more than orgasm. I've had disabled patients who are physically incapable of experiencing orgasms and still feel very sexual and enjoy a rockin' sex life. It's a state of mind. Own your sexual self, just as you are.

What's the deal with orgasms? Why are some strong while others are weak?

The strength of your orgasm depends on how much buildup leads to the orgasm, how aroused you are, and how much blood flow and oxygen reach the genital region. Also, the strength of the pelvic floor muscles, specifically the pubococcygeus (PC) muscles, directly affects orgasmic response. For example, many women describe orgasm during pregnancy as the best they've ever had. Why? All that extra blood courses through the genitals as the result of pregnancy hormones. And women who were taught to do Kegel exercises to improve urinary incontinence discovered, as a side benefit, that their orgasms increased in intensity.

Interested in strengthening your orgasm? Try Kegel exercises. Or talk to your doctor about the prescription-only Eros Clitoral Therapy device, a device that draws blood flow to the clitoris and is FDA-approved to treat female sexual dysfunction (although a variety of $19.95 sex toys may serve the same purpose and save you big bucks). Another tip from sex expert Lou Paget: Try panting and deep breathing while you're cresting to orgasm, which strengthens orgasmic intensity. Holding your breath does the opposite.

Does the size of my clitoris predict the strength of my orgasm?

No, it shouldn't. But having a larger clitoris can make it easier to find. It's the intensity of arousal and stimulation that determines the strength of the orgasm. But size does matter, in some ways. When you're aroused, the glans of the clitoris may disappear under the clitoral hood when blood-filled tissue swells. Your partner may wonder where you went. But if your clitoris is larger, it may be easier to find, especially during oral sex. And if you're one of the lucky ones who can orgasm during intercourse, a larger clitoris may be easier to stimulate. But size doesn't affect strength—just ease of access.

If I have a hysterectomy, will it affect my orgasm?

That's certainly a reasonable question. After all, if the uterus contracts during orgasm, how will its absence affect how we feel? The last thing you want when you lose your uterus is to also lose your O.

The good news is that most studies that assess sexual function after hysterectomy find that, if anything, sexual function tends to improve after surgery. It makes sense, when you think about it. After all, if you got a hysterectomy, chances are you got it because you were hemorrhaging, you had chronic pelvic pain, you had cancer, or big fibroid tumors were getting in the way (all pose barriers between you and sexual nirvana). Removing the source of what comes between you and sexual bliss may actually improve your love life.

In one study, the rate of sexual activity increased from 72

percent before surgery to 77 percent after surgery and the rate of pain during sex decreased from 19 percent to 4 percent after surgery. Before surgery, 92 percent of patients experienced orgasm, whereas 95 percent experienced orgasms after surgery.[10] However, losing the ability to orgasm is always a potential risk when you undergo hysterectomy because the nerves and blood vessels responsible for orgasm course right through the surgical field.

Keep in mind that in women sexual function is largely dictated by our minds. If you believe you are sexy and vibrant, you will likely remain sexy and vibrant after surgery. If you're convinced you are less of a woman after hysterectomy, you're more likely to experience sexual dysfunction.

So if you must say good-bye to your uterus, let go of your womb and throw your heart into embracing your sexuality, which is always yours to claim. Remember that your feminine spirit lies deep within you and can never be removed with any surgery.

If you are grieving the loss of your uterus and this is affecting your sex life, humor me and do this hysterectomy release ritual. Write a letter to your uterus on a sheet of paper, expressing all that you feel. Then dig a hole outside and bury the letter in the hole. Cover it with soil and plant a beautiful plant in the same hole. By releasing your uterus, you will watch as new life springs from the vessel you have lost. Your femininity and sexuality will blossom alongside the new growth. You just might find that your orgasm flowers alongside the plant.

Is it possible to have an orgasm mentally, without being touched at all?

Yes, it is definitely possible. Cherise had her first orgasm as a teenager, while watching a pornographic movie. Since then, she has experienced hundreds of orgasms using visualization alone. For her, masturbation requires no vibrators or clitoral stimulation. She can close her eyes, picture erotic imagery, and get off.

Dr. Beverly Whipple studied women who can orgasm by fantasy alone. The researchers discovered no significant difference between the orgasmic responses that resulted from imagery alone and orgasm that occurred as the result of masturbation.[11]

I keep hearing about tantric sex, but I don't really understand it. What is tantric sex and how can it benefit me?

While many Westerners link the idea of tantric sex with woo-woo workshops, multiple orgasms, and yet another way we're not quite sexually adequate, tantra is, at its essence, a type of yoga—the yoga of love. Pink Goddess of Mojo Caroline Muir, founder of the Divine Feminine Institute and co-author of *Tantra: The Art of Conscious Loving*, says,

> The "basics" of Tantra Yoga and its sexual component honors the female differently than it honors the male. Primarily . . . her pleasure, her awakening and her healing are paramount. Releasing her limitless pleasure becomes the focus of the Tantric lover. The man must learn to drink the energy of her massive power, thereby increasing his life force, or Yang energy. Tantra is ancient knowledge, complex in the spiritual/sexual realm. For modern Westerners, Tantra is thought to be primarily about sex. The "basics" include intimacy (connection), breath control, internal muscular control, and ways to deal with the often hidden material that lies within the female psyche once access to her soul-filled and limitless sexual power is attained.

The lovemaking becomes a kind of prayer, and lover and beloved make love as an offering to the divine within their own hearts. Without equal focus on the heart, on the love that dwells within, sex is simply scratching an itch, fulfilling a biological need. Nothing wrong with that at all, but it is not Tantra.

The man who gains mastery by holding off his orgasm (and subsequent ejaculation . . . requiring for most men a refractory period before his next erection) is one of the subjects important to becoming a Tantric lover. The man must learn to recirculate his life force or seed in order to build his sexual mastery. One of the sad truths about modern Tantra is how it has been simplified to be like fast food . . . quick, drive-through, satisfying . . . but not really fulfilling. At some point in the maturation process, there is a longing for deeper connection.

Many women do not have orgasm after orgasm even if the man is holding off his climax. Many women just get sore from all the friction, ending up with yeast infections, UTI's, broken hearts and a loss of hope. Many women are in need of sexual healing and awakening. I have taught tens of thousands of men and women the techniques for awakening the sexual power inherent in women, teaching men how to become sexual healers as well as tantric lovers . . . many women need to go through a period of very vulnerable and emotional content, revealing blocks to her energy for more pleasure. Many women have been emotionally or physically wounded (rape, unwanted pregnancies, abortions, incest, herpes, genital dis-ease, lack of self-esteem or body image). Sex becomes something women do for men until they just don't want to do it any more!

Please refer to www.sourcetantra.com for a complete education on Tantra Yoga. For at the core of sex in conscious, evolving humans, there needs to be an awareness in ALL of the centers of primary energy (power, heart, communication, positive imagery, spirit and groundedness . . . otherwise we are simply expressing our animalistic urge to copulate. That does not raise consciousness . . . that usually puts us to sleep. Tantric practices are ancient, designed to assist the lovers further and further into an awakened or enlightened state of awareness. It's not for every one, but primarily for those desirous of including their sexuality on their spiritual path.

So you see, the "basics" are beyond the physical and into the metaphysical . . . that is, if a raising of consciousness is desired.

Sounds good to me. Sign me up!

CHAPTER 6

Discharge and Itching

VAGINAS CAN BE MILKY, STICKY, slippery, clumpy, fishy, runny, itchy, and scratchy. They can flow, spout, pour, burn, and blush. It just happens. Vaginas have minds of their own, and sometimes they get a little out of control. Don't get mad at the poor vagina. She can't help it. Instead, listen to what she's trying to tell you. You may be surprised by how wise she is.

Valerie knows this well. She was a frequent flyer in my practice. At least once a month, she showed up with coochie complaints—itching, burning, irritation, and discharge. Most of the time, I examined her and found only a healthy, pink vagina. She would shake her head, insisting that I was missing something—a yeast, a bacteria, a parasite, something. I could do nothing but shrug. Finally, I asked her to keep a vagina diary. Each day, I wanted her to record how her vagina felt, so I could get a sense of her symptoms and try to figure out what was bugging her. I was thinking "Monday, itchy. Tuesday, burny. Wednesday, more discharge than usual."

A month later, Valerie came back, holding a pretty pink leather journal with a rose embroidered on the cover. Without saying a word, she handed me the journal. I opened it and read it.

Tuesday
Dear Valerie,
I feel like shit. I'm itchy, and this watery discharge is coming out of me, and to top it all off, you made me go see the doctor, who shoved that creepy metal thing up me. Right now, I'm really not feeling like talking to you at all.

Wednesday
Dear Valerie,
Sorry for being so testy yesterday. I'm feeling better today—a little less itchy, but I'm still way too wet. And I hate these panty liners you make me wear. Can't you find something softer?

Thursday
Dear Valerie,
I feel burny today, and honestly, I feel a little burned. I mean, what right does Bob have to talk to you that way? Why don't you learn to stand up to the bastard?

Friday
Dear Valerie,
I don't mean to keep complaining, but I'm feeling pissed off today. Bob keeps wanting to come inside for a visit, but

*I feel like he has no right to invade my privacy. He just hasn't
earned it lately. I'm putting my foot down and saying no.*

> *Saturday*
> *Dear Valerie,*
> *I'm having a good day. Actually, I feel like a million bucks.
> But I still don't want Bob to come visit.*

For the rest of the month, Valerie's vagina checked in with
her every day. The last entry read like this:

> *Dear Valerie,*
> *I know you're going to the gynecologist tomorrow, and
> she's going to want to see what I've written. I'm not sure this
> journal is what she had in mind, but tell her the itching and
> discharge aren't bugging me anymore. What's bugging me is
> you, always bitching about discharge and itching. I mean,
> seriously, lady. Get a life. Quit complaining about me all the
> time. Why can't you open your eyes. Bob is a schmuck, and
> you've put up with him long enough. Frankly, I can't stand the
> sight of him. He's been treating you like dirt for decades, and
> you just sit back and take it. Then you gripe about me, when
> he's the one you should be bad-mouthing. Listen, if you're not
> happy in your marriage, don't take it out on me. It's not my
> fault. Don't you think it's time you got some guts and changed
> your life, rather than moaning and groaning about me? I know
> you're scared. But don't worry. I'll be with you the whole time.
> Oh, and tell that gynecologist thank you for giving me a voice.*

When I looked up from the journal, I saw that Valerie was
weeping. When she was finally able to speak, she whispered,
"I'm leaving my husband."

A year went by before I saw her again. When she came in for a Pap smear, she looked like a completely different woman. Instead of leaning forward, with her shoulders hunched over, she stood up straight, tossing her newly styled hair and wearing a figure-flattering sundress. She positively radiated.

Valerie filled me in on the past year of her life. After our last visit, she went home and announced to her husband, Bob, that she was leaving him. Once she moved out, her coochie complaints disappeared. She continued the vagina diary, writing every day, and she finally realized that the itching and discharge had given her an excuse not to be intimate with her deadbeat husband. Once she no longer needed the excuse, her vagina was free to thrive. She shared a final passage with me from her pink journal:

Dear Valerie,

Victor is absolutely the best. I mean, I just blush thinking about how much time he spends admiring and nurturing me. I've never felt so worshipped in all my life. I get all moist just thinking about him. Thank you for listening to what I've had to say. Oh, and thank that nice gynecologist for encouraging us to chat. Tell her I'm sorry I haven't been nicer to her in the past. I'll try to cooperate next time she brings out the creepy metal duckbill.

What's the point of vaginal discharge? Why does it even exist?

I know vaginal discharge can seem like a pain. After all, it stains our panties, makes us feel sticky, and leaves little crusty things dried on our pubic hairs. So what's the point? Why must we endure such indignities? Keep in mind that just like the mouth and the nose, the vagina is a mucous membrane, serving as a doorway between the often dirty, germ-filled outside world and the body's precious, delicate interior. As such, the vagina has a job to do, and vaginal discharge is part of this ever-important function.

Vaginal discharge cleans and refreshes the vagina. When the fluid flows out, it takes with it old cells that line the vagina, making room for new, healthy cells that replace them. The contents of vaginal discharge also help the vagina retain a healthy acidic pH, which is necessary to fight infection and protect the vaginal environment from the outside world. It also serves to keep the skin moist. Without it, vaginal skin would dry up and become itchy and painful. So don't curse your vaginal discharge. Bless it. It's there for a reason.

What is vaginal discharge made of? It it toxic waste or something?

No, it's not toxic waste. In fact, unlike urine and feces, it doesn't contain any waste products. What it does contain is:

- **Fluid that seeps through the walls of the vagina**
- **Cervical mucus**
- **Uterine and tubal fluid**
- **Secretions from glands in the vulva**
- **Oil and sweat from vulvar glands**
- **Old cells from the walls of the vagina**
- **Healthy bacteria**

Your vaginal discharge consists mostly of salt water, mucus, and cells, things that normally exist in your body. There's really nothing icky about it. Vaginal discharge is to your tailpipe what a lube job is to your car. It helps maintain a healthy system and keeps your motor running.

I get crusty yellow stains on my panties, but I don't have any itching, burning, or odor. Is this normal?

Yes. Chances are, your crusties are normal. While some infections, such as gonorrhea and chlamydia, may present with vaginal discharge in the absence of itching, burning, or odor, it's normal to have a discharge, and when that discharge dries on your panties it tends to look yellowish and crusty. Just like the nose and mouth secrete fluid to stay healthy, the vagina does, too. In fact, if you put a panty shield over your open mouth for twelve hours, you'd likely find discharge, and it would probably be yellow and crusty. While most normal healthy discharge is whitish in color when wet, it may appear yellow when the fluid evaporates out of it. But if your vaginal discharge appears greenish when wet, you have itching or burning, your discharge smells extra-fishy, or you think you're at risk for STDs, get it checked, just to be on the safe side.

I'm a lesbian and my girlfriend often notices that there are chunky things stuck to my pubic hair, but this doesn't happen to her. It makes me self-conscious and I don' t know what to do.

Every woman makes different amounts of vaginal discharge. Crusties are simply vaginal discharge that clings to pubic hair and dries, similar to the "sleep" you may notice in the corners of your eyes when you first awaken. Because of your unique physiology, your girlfriend may make less vaginal discharge than you, or she may shower more often or clean herself periodically with wet wipes. Most women notice some crusties from time to time. If the crusties bother you because of sex, simply wipe yourself off before you plan to be intimate. Or consider trimming or getting rid of your pubes, which can catch vaginal discharge and lead to more crusties. But talk to your girlfriend. If she doesn't care and it's not bothering you, there's nothing wrong with a few crusties.

My friend says she barely makes any discharge at all, but I'm like a frigging faucet. How much discharge is normal?

Believe it or not, somebody actually studied this. Apparently, the average amount of vaginal discharge a woman of reproductive age secretes over a period of eight hours weighs 1.55 grams (a gram is equivalent to about 1/4 teaspoon). How much is normal varies depending on where you are in your cycle. You produce the greatest amount of discharge around the time of ovulation (1.96 grams).[1] Of course, because every woman is different, some produce very scant amounts of discharge and others make much more.

Is vaginal discharge the same thing as vaginal lubricant, or are they totally different?

Vaginal discharge is a catchall term that pretty much refers to all non-bloody juices that come out of Juicy Lucy. While vaginal discharge does help to lubricate the vagina on a daily basis, it differs from the vaginal lubrication you produce during sex. During the excitement phase of the sexual response cycle described by Masters and Johnson, the smooth muscles of vaginal blood vessels relax and fill with blood. More blood means more fullness, more responsiveness, and more lube. The lube comes from special ducts called Bartholin's glands, which are located right around the vaginal opening. In addition to making intercourse more comfortable, this lubrication helps keep sperm happy. Pretty nifty, eh?

Sometimes strange, grape-sized gobs of clear, gelatinous goo come out of my vagina. They look like pieces of a clear jellyfish. What is that?

My guess is that you're noticing these vaginal jellyfish around the time of ovulation, which is usually about fourteen days after the first day of your period. To facilitate transport of sperm through the cervix, cervical mucus changes around ovulation, becoming clear, goopy, and sticky, not unlike clear egg whites. Sometimes, if you haven't been drinking enough water and are a bit dehydrated, your cervical mucus can become even thicker around the time of ovulation, more like clear Jell-O or nose boogers. Either way, it's nothing to get worried about. And if you're trying to get pregnant, this is your signal to go for it! If you are trying to avoid pregnancy, beware when the jellyfish shows up, as it tends to signal your fertile time.

When I start spewing discharge, how do I know
if it's a yeast infection or something
more serious?

If your vagina normally runs on the dry side and then suddenly you're spewing gunk, chances are good that something is off. Many assume that all itching or discharge equals yeast. As it turns out, this just isn't true. How can you tell the difference between a yeast infection and other causes of abnormal discharge? Usually, a yeast infection causes a classic itchy, burning, red cooch, mixed with the signature clumpy, cottage-cheesy discharge. But other conditions can mimic these same symptoms.

Take Ellen, my patient who noticed itching, burning, and increased vaginal discharge. When I examined her, it turned out her symptoms had nothing to do with yeast. Instead, she was reacting to the thong underwear she started wearing while working out because she had a crush on her personal trainer and wanted to look sexy and sleek in her gym clothes. And then there's Mary, who thought she had persistent yeast infections, which turned out to be related to *vulvodynia,* a whole other condition that can mimic chronic yeast infections.

If you're clawing at your cooch, but you don't have a discharge, it could be any number of things: an allergy to your new tutti-frutti berry-scented soap, a dermatologic condition of the vulva, or waning estrogen levels, among other things. If you're spewing discharge without itching, it's more likely to be either bacterial vaginosis, a bacterial infection characterized by a foul-smelling discharge, or a sexually transmitted infection, such as gonorrhea, chlamydia, or trichomonas. If you can't tell the difference, see your gynecologist. It's pretty easy for us to help you sort it out.

What causes my yeast infections? Is it because I don't clean very well down there?

No. In fact, it may be that you clean *too* well. The vagina is supposed to be colonized with good bacteria like lactobacillus, which keep yeast at bay. But if you're douching, taking antibiotics, or scrubbing with antibacterial soaps, you may be killing the very bacteria that protect you. When the normal milieu of the vagina gets thrown out of whack, yeast can take over because the good bacteria aren't there to fend them off. If your immune system is weak, as it may be during pregnancy or with chronic health conditions, yeast may overgrow more easily. And yeast love sugar, so diabetics and those who eat foods loaded with sugar are more susceptible.

Where does the yeast come from? Usually, the yeast sprints over to the vagina from the anal area, hops on your maxi pad, or winds up in your vagina from your partner's mouth (it can go the other way, too). A few yeast buds in the vagina are normal. It's when they reproduce and go haywire that you wind up clawing at your cooch.

I've heard eating yogurt cures yeast infections. Is this true?

If you're gobbling up yogurt trying to treat your yeast infection, you're probably wasting your time. Yogurt does not contain enough lactobacillus to make much difference, and lactobacillus ingested this way does not seem to adhere to the vaginal walls. Also, because yogurt can be high in sugar, it may even make things worse. If you want to try it, make sure you choose yogurt without sugar. Probiotic supplements are probably a better way to get lactobacillus into your system.

When I have a yeast infection and I buy over-the-counter vaginal itching creams, they help the itching, but only while I'm using the cream. As soon as I stop, the itch comes back with a vengeance. What am I doing wrong?

Over-the-counter yeast creams can be tricky. Some of them are such skilled marketers that you'll think their product is the be-all-/end-all yeast cream. Unless you read the fine print, you may not realize that many of these creams target the symptoms of the yeast infection, such as vaginal and vulvar itching, but they don't actually cure the little yeasties that lie at the root of the problem. For example, Vagisil can be a soothing treatment for yeast symptoms, but it's not a cure like Monistat, which contains yeast-fighting anti-fungal medication. So if you're only using an anti-itch cream, your itching may improve, but your symptoms will likely return as soon as you stop it. In fact, your symptoms may be even worse, since the yeast will have had time to multiply and wreak havoc on your poor coochie.

If you've had yeast infections before, and you're pretty sure you know what's up down there, try an over-the-counter anti-fungal cream, such as Monistat, and take some probiotic supplements to help repopulate the vagina with healthy yeast-fighting bacteria. Then, if you like, you can use the anti-itching creams, with the confidence that you're already waging war against the underlying infection.

If I'm pretty sure I have a yeast infection, is it kosher to just pick up some Monistat at Walgreens and be done with it? Or do I have to take the day off work so I can see my gynecologist?

If you ask this question to a group of gynecologists, you're likely to elicit passionate responses. (Yes, we do get passionate about odd things.) Some think every woman with vaginal itching, discharge, or odor should be examined by a doctor because so many women misdiagnose themselves, putting themselves at risk of missing more serious conditions, such as chlamydia or a retained tampon that can lead to toxic shock syndrome. Others see no reason to bother a busy woman with an emergency gynecologic visit when she's likely to be able to take care of herself with a simple trip to the drugstore. Arguments between both sides may erupt (and you ain't seen nothin' till you've seen a gynecologist catfight!)

Personally, I'll go out on a limb and say that I think it's just fine to pick up that Monistat from the drugstore if you have the classic itching, burning, and clumpy white discharge that usually signal the presence of a yeast infection. If your symptoms resolve, you can rest assured, knowing that you most likely had a yeast infection, and you're probably good to go. If your symptoms aren't classic, you're not convinced it's a yeast infection, you're engaging in risky behavior such as unprotected sex, you're getting frequent yeast infections, or you try Monistat and don't get better, it's time to check in with your gynecologist.

The older I get, the dryer I feel. What's up down there?

The vagina's moisture level varies throughout your lifetime. When we are babies just born from mothers who have estrogen coursing through their pregnant bloodstreams, our girl parts appear plump, pink, and fleshy, responding to the estrogen we are exposed to in utero. Then, throughout childhood, our estrogen levels fall to negligible levels. During this time, the vagina is reddish in color, very fragile, and extremely dry. Then puberty hits and—wham!—estrogen reappears, making the vagina fleshy again and stimulating secretions that lead to vaginal moisture. Moisture peaks during pregnancy, when estrogen levels surge, and falls off temporarily in the postpartum period for mothers who nurse, leading to temporary vaginal dryness. Once we stop breast-feeding, the vagina regains its moisture. As we age, estrogen levels fall off and, unless we treat our bodies with hormones, the vagina dries out.

If vaginal dryness is bothering you, over-the-counter vaginal moisturizers like Replens may help. If your dryness is getting in the way of your sex life, talk to your lover about using saliva to lube you up. Not only will you enjoy the benefits of a little licking, but this natural lubricant can prevent chafing and pain that comes with vaginal dryness also. If salivary lube fails to do the trick, try using a sexual lubricant like K-Y Jelly or Astroglide. Or experiment with jojoba oil or vitamin E on the vulva. If that doesn't do it for you, vaginal or oral estrogen replacement can do wonders to get your vagina in tip-top shape.

CHAPTER 7

Periods

WHEN I HIT PUBERTY, I counted the days until my period began, just like all the other fans of Judy Blume's coming-of-age story *Are You There God? It's Me, Margaret.* Every morning, I hunted for those precious drops of blood, but again and again, I found nothing in my panties.

It seemed like every girl at school got her period before I did. "God, I hate it when Aunt Flow comes to visit," said bouncy Jennifer as she pranced around the locker room in her lacy lavender bikini panties that bulged in the crotch from her maxi pad. Jennifer was one of the cheerleaders who wore short little skirts to school on Fridays and who had already kissed multiple boys.

"Aunt Flow?" I asked.

"Yeah . . . you know, when *Aunt Flow* comes?"

I was clueless. Another girl laughed at me. "Come on, Lissa. You know. *When Aunt Flow comes to town?*" She snickered. "Does Flow not come to your house?"

I shook my head. "I don't have an Aunt Flow." The girls laughed and whispered.

Of course, as luck would have it, Aunt Flow finally came one Thursday morning, right at the end of English class, when I was wearing a white sundress and didn't have a maxi pad. Feeling an unfamiliar moist, drippy sensation, I dashed out of the class-room, scuttling sideways like a crab with my back toward the row of lockers until I reached the girls' bathroom. There it was: my scarlet letter. A crimson bloom right across the back of my dress. At the sink, I found Cindy, a girl I barely knew, and whis-pered, "I have an emergency!" I turned around quickly, pointing to the stain on my dress, then spun back around as fast as I could.

"Lissa!" she exclaimed. "Aunt Flow finally came!" *Oh great,* I thought. *Everyone* knows *I've been waiting for this.* But I breathed a sigh of relief. Cindy linked her arm in mine, threw her sweater around my waist, and led me to her locker. She grabbed a clean T-shirt and a pair of shorts, along with a little pink cloth zipper case, from her locker. "I'm always prepared," she said. I felt a rush of warmth from the girl-bonding and vowed to always be on the lookout for anyone who needed a helping hand in the fu-ture. Little did I know how many women with menstrual prob-lems I would wind up helping.

I wore the maxi pad Cindy gave me for exactly four hours and twenty-two minutes, counting down the seconds until school ended. I had never felt so disgusted in my entire life. My crotch was wet and soggy, and I could feel fluid dripping out of me and seeping into the diaper that squished between my legs.

When I got home, I told my mom I had started my period. "Great, honey," she said. "You're a woman now."

Yippee. Woo-hoo. Yay for me. After I surrendered to the experience of letting my mother put in my first tampon (long story—I'll spare you), more menstrual nightmares awaited me. One night, when my parents were throwing a party, our mischievous bichon frise puppy found my used tampons in a wastebasket. With my blood staining the fur of her mouth and nose, she pranced around and dropped the soiled cotton at the feet of our party guests like they were chew toys, then begged to be petted. Everyone in the room was noticeably uncomfortable, and I wanted to simply disappear. Finally, one of our guests broke the silence.

"You should have seen my mother's face when I was a teenager and my dog dragged out all of the condoms I had used with my girlfriend and then carefully hidden in the trash," he shared. People smirked awkwardly, but it did nothing to draw attention away from my blood-soaked dog, who by this time was pulling out all her party tricks in order to get attention. Suffice it to say I was seriously scarred.

By the time I was in my twenties, I sent my uterus on a ten-year sabbatical. With the manipulation of modern medicine, I went more than a decade without menstruating. During my medical training, I felt like my menses were the only barrier keeping me from doing everything a man could do. I refused to be held back and swore off menstruation forever.

Now, at forty, after having a Mirena IUD inserted right after giving birth, I am once again free of the monthly reminder of my femininity. But I wonder—by chemically manipulating my body to rid me of the hassles of menstruation and avoid pregnancy am I missing something, some sacred cycle of life that would make me more whole? Am I sacrificing the richness of the full female experience for the sake of convenience? While there's no medical reason for me to have a period, might I somehow be denying the Goddess within? Maybe the sacred feminine within me has things to say and I'm too busy to listen.

I know I'm not alone in having a complex love-hate relationship with my yoni and my menstrual cycle. Most of my patients have similarly confused feelings. While I don't claim to have the answers to all things gyno-spiritual, I can help you clear up a few of the misconceptions that tend to surround this mysterious process.

Why do women have to deal with the misery of getting periods?

Periods might seem like a curse. It's tempting to blame poor Eve for what we women endure. But really, menstrual cycles are a blessing. Just ask Ellie, my twenty-five-year-old patient who stopped getting periods and can't have babies. She'll remind you that your period is a regular sign of your reproductive potential.

But why do we get periods? Every month, your body gets ready to have a baby. The hormones in your brain signal to your ovary that it's time to start preparing an egg. One lucky egg wins out as the dominant follicle and gets released. The uterus prepares a soft, spongy blood nest in anticipation of the embryo it hopes to support, and the lining of the uterus thickens. Once the body discovers that the egg has not been fertilized and there's no pregnancy, it sheds the lining and you menstruate. Then, ever hopeful that it might fulfill its biological destiny, the body starts all over again by preparing for the next month.

Why do I get clots when I get my period? Should I be worried?

Some women get all wigged out about passing blood clots, but actually, clots are good news.

They mean that all of your clotting factors are doing exactly what they're supposed to do when you bleed. I've had patients fish clots out of the toilet, put them in little Tupperware containers, and bring them in to show me. While I appreciate the, um, gift, rest assured that clots are nothing to worry about.

All a clot means is that the blood has pooled somewhere and coagulated before coming out. For example, it's not uncommon for blood to pool in the vagina when you're sleeping at night. The blood clots, then voilà! When you wake up and go to the bathroom, clots come out. Also, what may look like clots are often clumps of uterine lining you have naturally shed, rather than clotted blood. Sometimes the two can mix together into scary-looking (but usually harmless) goombahs.

What matters more to me, as a gynecologist, than clots is the amount of blood you lose during a period. Liquid blood pouring out of a vagina like a soda foundation spewing cherry soda makes me more nervous than clots on a maxi pad. Sometimes, but not always, clots may mean you're bleeding more heavily than usual. I care how big the clots are. Are they the size of a blueberry? A grape? A strawberry? A lemon? An orange? A grapefruit? A canteloupe? A weeklong period full of strawberries could be a problem. By the time we're talking about citrus fruits and melons, I'm definitely worried.

If you're bleeding heavily, with or without clots, call your doctor right away. If you're also dizzy, light-headed, pale, or passing out, get thee to a hospital. But don't stress about the occasional clot. It's just nature doing its job.

I have abnormally heavy periods. I go through a super-absorbency tampon and a giant pad in less than an hour. What's wrong with me? Do I have a giant uterus or overactive hormones or something?

If you're filling both a super-absorbency tampon and a giant pad in less than an hour and that's happening hour after hour, day after day, something is definitely wrong. You may have the occasional heavy hour during a normal period. But if you're going through a box of tampons every day or you have to wear double protection to keep from soaking your clothes, you could have a problem. If you're bleeding heavily on a regular basis, chances are good that you're also anemic (low red blood cell count).

Why is that? Do you have a giant uterus? Maybe. Hormone fluctuations? Perhaps. If you're a young adult, a common cause of heavy bleeding is an unsuspected pregnancy in the process of miscarrying. If you're in your forties, common causes of heavy bleeding include fibroids (benign smooth muscle tumors of the uterus), uterine polyps (polyps lining the inside of the uterus), adenomyosis (when the uterine lining invades into the muscle of the uterus), the presence of a Copper T IUD, and hormonal imbalances. If you're bleeding heavily, please see your doctor.

While cancer is a relatively uncommon cause of heavy bleeding, it's a possibility you definitely don't want to miss. And you don't want to put your health in danger by ignoring life-threatening anemia. Be safe.

If I get my period when I'm scheduled for an appointment with the gynecologist, should I cancel?

It depends on why you made the appointment in the first place. If you're scheduled to see your gynecologist to chat about birth control options, it really doesn't matter whether you're bleeding. In fact, if you scheduled your appointment for any reason other than a Pap smear or procedures such as colposcopy or hysteroscopy, go ahead and keep your appointment.

My patients get embarrassed when they're bleeding, as if I'll be grossed out by a bloody tampon. But this is what we do all day. We deliver babies. We operate. We look at blood. We honestly don't give a flip whether you're on the rag. And if you're there because of heavy or irregular bleeding, you're actually helping us better assess the whole picture.

If, on the other hand, you're scheduled for your annual exam and Pap smear, you might consider rescheduling if it's not inconvenient. First of all, you might feel more tender when you're menstruating, so a Pap and pelvic may be less uncomfortable when your period is over. Also, if there are too many red blood cells present when the pathologist goes to read your Pap smear, they may have a hard time sorting through the blood cells to see the cells from the cervix. Worst-case scenario, your Pap smear will need to be repeated.

If my patient is scheduled for a Pap and happens to come in when she has her period, I'm happy to do her Pap anyway. I use a big Q-tip to wipe off extra blood, Pap gently, and use a light touch when examining her. Most of the time, the pathologists can read the Pap just fine. Most of my patients prefer that to having to reschedule, since it's not always easy to rebook appointments quickly. If you're uncertain of your doctor's policies, just call and ask.

I despise getting my period and my doctor says I can skip it with birth control pills. It just doesn't seem natural. Is it safe not to menstruate? What happens to all that blood?

While skipping periods when you're not taking birth control can signal trouble, skipping your periods while taking continuous birth control is a whole other ball game. For example, if you take birth control pills and, instead of taking the week's worth of sugar pills that cause you to have a period, you skip the sugar pills and open another pill pack, you will likely skip your period. If your body lets you, you may be able to do this indefinitely.

Is it safe to skip your periods with birth control? Yes. While birth control pills are not without risk, taking birth control continuously does not seem to add additional risk. When you skip periods while taking continuous birth control pills, it's not as if the uterine lining is building up and not shedding. Instead, the uterine lining becomes atrophic, meaning that it simply stops growing. In essence, there's nothing to shed.

Is this natural? No! Of course not. It's natural to get pregnant a dozen times in your lifetime. There's nothing natural about hormonal birth control. We defy nature when we choose to take hormones to prevent pregnancy. But natural or not, we consider it safe to skip your periods using hormonal birth control under a doctor's guidance. In truth, the fact that you menstruate on a regular birth control pill cycle is completely arbitrary. The only reason pharmaceutical companies manufactured pills this way is because many women prefer to get a period, either to ensure that they are not pregnant or because they assume it benefits their health to menstruate.

When you inform women on the Pill, patch, or vaginal ring that they need not menstruate, only about 30 percent would prefer to menstruate monthly.[1] Would you like to skip periods? Talk to your doctor about the risks and benefits and whether this might be an option for you.

I always get horny during my period. Is it unusual to crave sex during your period?

No, it's not unusual. My patient Drew felt the same way.

"I'm like a dog in heat," she said, blushing. "It doesn't make any sense to me. I get it for dogs. When they bleed, that's their fertile time, so it would make sense that the dog would be horny when she's in heat. But what's the deal with me?"

She has a good point. Humans aren't usually fertile during their periods, so why do some menstruating women want to jump any guy who comes within a six-foot radius?

Drew was too embarrassed to tell her husband how much she wanted to make love with him during her period. He assumed she would prefer to avoid sex when she was bleeding, and she figured he would be repulsed by her menstrual blood. One day, he walked into her bedroom when she was masturbating during her period. Stimulated by what he saw, he stripped off his clothes, and they went at it like bunnies—and have been at it ever since.

On the flip side, Heather said her boyfriend couldn't resist her when she was menstruating. "It's like I'm secreting pheromones," she said. But sex is the furthest thing from Heather's mind during her period. She feels crampy, bloated, and withdrawn during her period. Her whole pelvis swells, and any little touch to her genitals feels unpleasantly sensitive. She craves Motrin, chocolate, and a heating pad during her period, but definitely not sex. Drew and Heather represent opposite ends of the spectrum with regards to menstruation and sex, but both are perfectly normal.

Why do some women notice a super-charged sex drive during menstruation? No one really knows. You can blame hormonal fluctuations, or it may just be that you're less scared of getting pregnant. Maybe relief from PMS gets you in the mood, or maybe the additional slip-slidey lubricated feeling juices you up. If sex is the last thing on your mind when you're menstruating, that's normal, too. It's okay either way.

Is it safe to have sex during your period?

If you're going to do the deed when you're menstruating, keep in mind that sexually transmitted infections may be more easily transmitted when there's blood involved. You also can't trust that you're not fertile during this time, so it's not safe to count your period as birth control. Occasionally, women do ovulate while they are still bleeding. Condoms can decrease your risk of both.

If you're not worried about pregnancy or infection, you're safe to go at it. If you do plan to have sex during your period, give yourself permission to be messy. Cover your bed with a washable waterproof cover, or simply use a towel. You can also try the Instead Softcup, a menstrual cup that collects blood, as an alternative to a tampon or pad. Be forewarned that this is *not* a contraceptive, but it can cut down on the mess if you engage in sexual activity during your menses. With your sheets safely protected, why not experiment with other messy stuff—personal lubricants, whipped cream, or coconut oil (but don't blame me if you find yourself jonesing for a postcoital macaroon).

Is it bad if a guy wants to go down on you while you're on your period? Will it hurt him?

Any time you exchange body fluids, especially blood, you have to consider the risk of infection. If your partner gives you oral sex when you're bleeding and has a tiny open mouth sore, infections can spread. If you're not worried about infection and he's game, the blood shouldn't hurt him. To reduce bleeding during oral sex, consider using a diaphragm, which can trap some of the menstrual blood until you're done having fun. You can also put in a tampon during oral sex, but make sure you take it out if you switch to vaginal intercourse. Many couples find creative ways to keep their sex life alive during menses. I say more power to 'em!

How come the blood coming out during my period is brown sometimes?

Some of my patients freak out when menstrual blood turns brown. But don't worry. Your blood isn't transmutating into poo. It's just getting old. When blood pours quickly out of your uterine blood vessels, it is bright red. (Trust me; I've seen this way too often.) When it sits around in the uterus before coming out, the iron in the hemoglobin molecules in the blood begin to break down. In essence, the blood begins to "rust." Generally, brown blood comes out of the uterus when bleeding is very light. It's no cause for concern at all.

Why do I get so hungry before and during my period?

Feeling hungry or craving sweets and other carbohydrates before and during your period most likely relates to hormonal fluctuations that make you more sensitive to your body's natural insulin. If your blood sugar drops precipitously, it triggers hunger and carb cravings. If you then eat sweets or simple carbohydrates to curb the cravings, your blood sugar shoots up and triggers your pancreas to spit out more insulin, thereby jacking up your insulin levels. And thus, the cycle continues. The best way to avoid these hunger pangs and carb cravings is to eat a diet that keeps your blood sugar steady. Vegetables, lean proteins, and whole grains keep your blood sugar and insulin levels stable and may dramatically reduce these symptoms.

*I forgot about my tampon and left it in for almost
two days. Will I get toxic shock syndrome?
When should I worry?*

While it's never a good idea to leave a tampon in for an extended period of time, your risk of toxic shock syndrome (TSS)
is still very low. In the early 1980s, the incidence of TSS rose to
approximately ten cases per one hundred thousand women. After educational efforts and tampon modifications, rates fell to
about one case per one hundred thousand women. Recently, for
reasons that are unclear, the incidence appears to be on the rise
again. But it is still very rare. Contrary to popular belief, most
cases of TSS are not due to a tampon stagnating inside the vagina for extended periods of time. Most women with TSS are
wearing a tampon and menstruating at the time of diagnosis,
but the majority of these women have changed their tampon
within a reasonable time frame.

Of course, this isn't meant to encourage anyone to leave
tampons hanging out in their vagina for days at a time. While
the risk of TSS may be low, it's still pretty gross. To play it safe,
never leave a tampon in longer than a good night's sleep.

*I felt a big lump at the end of my vagina when I
tried to get out my tampon? Is it cancer?*

Probably not. I vividly recall the day Eloise came racing into my
office without an appointment. She was so distraught that my
nurse put her into a room right away. By the time I entered the
exam room, she was hyperventilating and sweating, with a racing pulse.

She started talking so fast I couldn't understand her. When she finally calmed down, she said, "I think I have cancer."

As it turns out, Eloise was menstruating and couldn't remember whether she had removed her tampon. Remembering what I advised the last time she lost a tampon, Eloise went hunting. When one finger revealed nothing, she put two fingers in. Still finding nothing, she managed to insert her whole hand inside her vagina, and when she got to the top of her vagina she found no tampon. Instead, she found the mass that had prompted her emergency visit to the gynecologist. When I inspected Eloise's vagina, I saw only a healthy vagina and a normal-appearing cervix. The bump Eloise thought was cancer was actually her cervix.

If you're masturbating and you come across a bump in the vagina, it could be a wart, a cyst in the vagina, or (God forbid) a cancer. But chances are, if it's a big bump at the tail end of your vagina, which feels like your nose with a little hole in it, you're just feeling your cervix. If you're not sure or you're overdue for a Pap smear, see your doctor.

I swear, PMS turns me from Dr. Jekyll into Mr. Hyde. Why do I turn into a bloodletting, wild-eyed, raging lunatic before I get my period?

Studies have failed to elucidate a clear answer about what causes PMS. Most blame the fluctuations in our hormones and brain chemicals that happen during menstruation. But the truth is we just don't know for sure (and believe me, we know-it-all doctors hate to utter those words).

Some suggest that PMS exists because we expect it to exist

and those who grow up believing that they will experience PMS do. One study found that when you trick a woman into thinking she will menstruate at a different time, she reports PMS symptoms just before she thinks she will menstruate, not before she actually bleeds.[2] But try floating this theory to someone in the throes of crying spells and chocolate cravings. You're likely to wind up with a Kit Kat in your ear.

In her book *Women's Bodies, Women's Wisdom*, Christiane Northrup, M.D., suggests that we embrace the changes that happen with our hormonal cycles. Rather than cursing our hormones, she encourages us to rejoice in them, listen to what they tell us, and respond to their call. In the first or follicular phase of the menstrual cycle (days 1–14, where day 1 is the first day of bleeding, so usually the week of your period and the following week), creative energy is high and we are outgoing and attractive to others. New projects or ideas are best initiated during this phase. At ovulation, creativity peaks and we are receptive to others, a time we can optimize. After ovulation, in the weeks leading up to menses, progesterone dominates, as we listen and revel in contemplative reflection upon our creations and challenges. During this time, Dr. Northrup says, "women are most in tune with their inner knowing and with what isn't working in their lives."

Premenstrually, in the phase commonly known as PMS, Dr. Northrup claims we are more intuitive, more susceptible to tearful emotion, more angry, more connected to our pain, and more in touch with old problems. She suggests we accept this as a normal part of womanhood, rather than deny it, honoring our natural cyclic ebb and flow, allowing ourselves this reflective time, and using our menses as a period of rest and rejuvenation.[3]

Maybe we're not supposed to feel cheerful, productive, and at top performance of our game every minute of every day. In tribal societies, menstruating women were often isolated from the tribe, spending the time of their menses with other menstruating women, since blood was believed to attract predators and put the tribe at risk. I'll bet those women had a really good time in the Red Tent, temporarily relieved of their duties and bonding with other women. If we could honor our bodies, listen to our hormones, and put life on hold, perhaps our premenstrual slowdown could help us get in touch with who we really are.

But I hear you, ladies. If you're an attorney and a critical trial date is scheduled for the week before your period, you may be thinking, *Good luck getting a judge to understand my primal need for rest.* Indeed, we live in the modern world, but being attentive to our natural cycles can help us accept what is.

Do you have any natural tips for helping my PMS?

Yes. When my patients at the integrative medicine center where I work suffer from PMS, I recommend the following:

1. Eat a whole-foods diet. You've heard it before, and that's because it really does help. A whole-foods diet means cutting back on sugar, refined carbohydrates, caffeine, processed foods, and saturated and hydrogenated (trans) fats. Instead, add fruits, veggies, lean proteins, and whole grains, especially during the luteal phase (second half) of your cycle. And I'm so

sorry to break it to you, but that means bye-bye chocolate. Also, some women manifest gluten sensitivity and lactose intolerance with PMS symptoms, so if you're suffering, experiment with cutting out wheat and dairy.

2. Increase essential fatty acids by eating nuts, seeds, and fish.

3. Manage your stress effectively. Try yoga, meditation, massage, or guided imagery CDs.

4. Engage in regular aerobic exercise. A good, long hike helps regulate your hormones and your stress, and it builds up happy-making endorphins.

5. Take a multivitamin and twelve hundred milligrams of calcium a day.

6. Talk to an integrative medicine doctor about how supplements may help your PMS symptoms. Alternative medicine therapies like acupuncture and Reiki may also help.

Why do I get such bad intestinal cramps and diarrhea at the same time that I get my period? Isn't having disgusting glop coming out of one end bad enough?

Prostaglandins are chemicals that are released during the menstrual cycle and affect smooth muscle in the uterus, as well as smooth muscle in the bowel. These prostaglandins trigger menstrual cramping in order to push the blood out of your uterus and down through your vagina. At the same time, all these extra prostaglandins have a similar effect on the bowel, which is why intestinal cramping and diarrhea can often coexist with uterine cramping.

CHAPTER 8

Fertility

FROM THE TIME I WAS young, I cursed my uterus. Cramps plagued me when I was trying to do rounds at the hospital, and blood would leak out of my tampons and onto my scrubs in the middle of a surgery. Seeking a way to escape my own woman-hood, I discovered that I could take birth control pills daily and never get a period. Why hadn't anyone ever told me this? After I uncovered this secret, I sent my uterus to a dark recess of some basement closet and didn't bleed again for a decade. Every now and then, my uterus (I affectionately call her Yoni) would cry out for me, but I pretty much ignored her. I wasn't a very good friend.

Around the time I turned thirty-four, I heard Yoni calling more consistently, beckoning like a siren bellowing out to sea.

She'd cry, "Lissa! Lissa! Don't forget about me."

And I'd shrug her off: "No, Yoni. I'm busy."

She kept asking, "Aren't we ever going to have a baby?"

I responded with my standard brush-off answer: "Not now, maybe tomorrow."

As my thirty-fifth birthday loomed, I decided to bring it up with Matt, the commitment-phobic perpetual bachelor I had been dating for almost two years. But I wasn't quite sure how to broach the topic. Do you say, "So I've been chatting with my uterus lately"? Or do you couch it in the awkward terms of biological clocks and such?

Now, mind you, I was never one of those women who *had* to be a mother. I was always a bit on the fence. I love children, and with the right guy I could see myself having oodles of love to offer a child. But I didn't see myself trying to wrangle some guy into fathering my child if he didn't want to be a fifty-fifty parent. So it seemed a strange turn of events to find myself bringing it up. I guess I knew deep down that my beloved Matt would be a wonderful father, and I feared he'd realize this when I was forty-two and past my prime.

So I swallowed my pride and, heart beating fast, I broached the baby conversation. I have to give him credit. In spite of how shocked and blindsided he felt, he heard me, validated me, and promised to give it some thought. I didn't blame him for being surprised. After all, we had joked that we were the perfect couple. I was the twice-divorced girl who never needed to marry again, and he was the perpetual bachelor. But alas, things change.

I asked him to consider my question over the course of the next year—no pressure. I wasn't attached to any answer, and I wouldn't take it personally if he said no. But if, at the end of the year, he said no to having a baby with me, I wanted permission to take it off the table permanently. Not to judge anyone else's choice, but personally, I just didn't want to be a forty-something-year-old woman making a test tube baby because I had delayed childbearing too long.

Almost exactly a year to the date after that fateful day, Matt said yes.

I felt this rush of something—adrenaline, I guess—move through me, and I broke into a cold sweat.

I said, "Well okay then." Matt's face registered panic. Turns out he thought he'd have to do more convincing.

Almost as an afterthought, Matt asked awkwardly, "So, uh, should we get married, then?" No ring. No grand gestures. I agreed. We wanted our baby to know she was chosen.

Soon afterward, I took my last birth control pill and invited Yoni out of the closet.

The next month, Matt and I said our vows in a private ceremony in Big Sur. We fantasized about creating a honeymoon baby, but thirteen days after our wedding my period arrived, splendid in her red dress, leaving me curled up like the fetus I'd hoped I could create. In the throes of terrible cramps and faucets of blood pouring on my sheets, I cried, "Yoni! What are you doing?"

She said, 'Ahhh... I'm back."

To Yoni, I yelled, "No wonder I locked you up!" and to Matt, "Get me pregnant, *now!*"

Let's just say it was a confusing, exhilarating, and surreal time. I know I'm not the only woman who feels this way about her fertility. For many of us, fertility can be a hot-button trigger. Sure, our bodies signal fertility readiness when we're in eighth grade, but many of us aren't emotionally ready until we're pushing forty.

More and more, we delay even thinking about childbirth

until our ticking biological clocks turn up the volume. In my opinion, something is seriously wrong with evolution. With advances in women's rights, greater professional opportunities for women, and an epidemic of commitment-phobic guys, many of us simply aren't ready to reproduce in our twenties, when we're most fertile. And if you're one of those women in your late thirties or forties who are still waiting for Mr. Right, you may be in tears by now. It's just not fair. Why must women fit fertility into such a tight schedule? I mean, seriously. Not to question the Divine, but couldn't we work on fixing this little kink in the system?

If your fertility elicits strong feelings, you're not alone. Maybe you found yourself a young mom, long before you were ready, and now it's time for you to spread your own wings. Maybe you put off childbearing in order to pursue your career and now it's too late. Maybe you got pregnant when you weren't ready and chose to terminate the pregnancy. Maybe you got pregnant but lost your baby. Maybe you popped out your babies right on schedule, but you're so caught up in being a mother that you've forgotten who you really are.

Regardless of your relationship with your own fertility, spend some time with yourself to get in touch with the quiet inner voice of your heart. Instead of second-guessing your fertility decisions and yourself, listen to the guru that lies within you. Keep in mind that every decision we make informs who we are today. It's easy to slide into feelings of regret, but how can we regret what made us who we are?

Is there a test that can tell me whether I'm fertile?

If you've put off pregnancy to pursue your career, waited decades for the right partner, or delayed childbearing to get ready emotionally, you're just like many of my patients. Then one fateful day, you meet the perfect lover, your career is on autopilot, and you realize you're ready. But wait! Is it too late? Wouldn't it be great if you could simply take a blood test that would predict whether you're still fertile?

Unfortunately, it's not that simple. While a blood test called FSH (follicle-stimulating hormone) can be ordered by your doctor and FSH can also be measured more crudely by the over-the-counter First Response fertility test, these tests are far from perfect. While a high FSH level can suggest that your eggs are less fertile, a low FSH does not in any way guarantee that you will get pregnant. And I've seen women with menopausal-level FSH tests conceive. So you just never know until you try. If you conceive and give birth to a child, you're fertile. While you can undergo an extensive battery of tests to evaluate why you might be infertile, normal testing does not prove that you *are* fertile. And even those tests are imperfect. I've seen many women with abnormal testing go on to conceive naturally after years of expensive fertility treatment. When it comes to a woman's ability to conceive, I've learned to never say never.

We've been trying to get pregnant and I'm so impatient. Waiting is the hardest part. On average, how long does it take for a couple to get pregnant?

I know how frustrating it can be to wait. Once you've finally decided to take the plunge and stop your birth control, you're ready to start decorating the nursery already. When you're trying to plan for a baby, you end up putting your entire life on hold. Can you plan that ski trip over Christmas break? Well, it depends whether you're pregnant by then. Should you buy that fitted blouse that will certainly be out of season by next year? Hard to say—you might be wearing maternity clothes. Should you take that account that will culminate in a big presentation next October? Who knows. You might be on maternity leave by then. Waiting is so hard. And each month that goes by without a pregnancy signals yet another failure—and even more uncertainty. I feel you, sister.

In truth, it takes mere seconds to get pregnant. When just the right egg meets the perfectly suited sperm, they merge in moments. But how long does it take for those ideal conditions to align? It varies widely from couple to couple. If you're having sex at the right time, the chance of getting pregnant the first cycle you try is 30 percent. The second month, you have another 30 percent shot.[1] Most couples who are having sex around the time of ovulation conceive within six months. After that, the pregnancy rates per cycle seem to go down.[2] Studies show that 85 percent of couples get pregnant within a year. By definition, we label the 15 percent of couples who do not conceive in the first year as "infertile," although I hate that term. It feels so defeatist. I prefer the term *fertilely challenged.*

Over the next thirty-six months, about 50 percent of the remaining couples will conceive spontaneously. The 5 to 7 percent of couples who do not conceive spontaneously within two years are statistically unlikely to do so without fertility treatment.[3] But again, you just never know. I've seen it happen....

Keep in mind that the last thing you want to do when you're trying to get pregnant is stress about it. Stress impairs your fertility. (I know, I know. Now we need to stress about stress!) While waiting is hard, the best thing you can do for your fertility is have sex, have fun, and just go with the flow. If you're desperate to conceive, you're under thirty-five, and a year has passed, or if you're over thirty-five and you've waited six months, you might want to consider seeing a fertility specialist.

If I'm trying to conceive, does having an orgasm help?

Mary Roach, author of *Bonk*, reports that in the early 1900s doctors recommended orgasm as a treatment for infertility. She says, "The upsuck theory holds that the contractions of a woman's orgasm serve to suck the semen up through the cervix and deliver it quickly to the egg, thus upping the odds of conception. There is evidence that this is true with certain animal species. (Pigs, for instance. In Denmark, pig inseminators actually stimulate the sows while inseminating them for just this reason.) Noted sex researchers Masters and Johnson were 'upsuck' skeptics and designed a study to test this theory. By mixing simulated semen with radiopaque dye and placing it in a cervical cap, they could x-ray women during orgasm to see whether it was being sucked in by the uterus. They found no evidence that it was. Roy Levin, a retired sexual physiologist in the UK, points out that sperm take some time to capacitate, and thus you wouldn't want to deliver the sperm too quickly to the egg, because they're not up to the task yet. So the jury's still out. Unless you are a sow. Then yes, it seems to help."

On the other hand, another study found that women who have orgasms during intercourse after the prospective daddy ejaculates retained more sperm than either those who didn't orgasm or those who orgasmed before their partner.[4] And those who retain sperm in the vagina for ten to fifteen minutes may be associated with higher rates of fertilization.[5]

So it's hard to say. There's certainly no airtight evidence either way, so I say orgasm if you feel like it, but don't make it yet another thing to add to your fertility to-do list. Trying to conceive can be stressful enough, without heaping on additional expectations.

Can abortion cause infertility?

Not usually. But I've been asked that question by hundreds of tearful women. Those who choose to abort babies often have open wounds in their psyche related to their decision. It's an emotionally charged issue already, but if you add infertility to the mix, it's like salt on a wound.

Although I haven't had an abortion, I have stood beside so many women going through the procedure that I almost feel like I've experienced it personally. You find yourself pregnant from the wrong guy at the wrong time, and you do the best you can to make the right choice. You cry your way through the abortion, knowing you don't have any other choice at that moment in time. You heal and move on with your life. Then, ten years later, you've met the perfect partner, you've aligned your life to receive the gift of a child, but you can't get pregnant. Next thing you know, you're kicking yourself, wondering if that long-ago pregnancy was your one-and-only chance to be a mother.

Having trouble getting pregnant often triggers intense guilt, self-hatred, and doubt in women who have previously chosen to terminate a pregnancy. If you're one of those women, please, give yourself a hug. What's done is done. It doesn't benefit you to beat yourself up about a decision you can't undo.

While you may be tempted to slide into a free fall of "what ifs," this won't help you one bit. You can only truly live in the present moment. But I'll get off my gynechiatrist soapbox now.

Rest assured that your fertility challenges are likely unrelated to your past choices. Unless you suffer serious-but-rare complications from your abortion, such as pelvic infection or scar tissue inside the uterus, abortion usually does not affect future fertility.

My patient Elaine had two abortions as a teen, and when she tried to get pregnant at thirty-seven, she couldn't. She blamed her abortions and felt that God was punishing her for giving up her pregnancies. In actuality, Elaine suffered premature ovarian failure, meaning that her eggs behaved as if she were in menopause, a condition completely unrelated to her abortions. Personally, I believe in a loving God who feels compassion for the pain Elaine is experiencing, not a vindictive deity who strips women of their fertility out of spite. But that's just me....

So please. Be kind to yourself. It's in the past, and there's no point regretting something from your past. Instead, explore the life lessons you might learn and appreciate the fact that all of our experiences, even the painful ones, help us evolve and grow into the people we are meant to become.

I've heard that abortion increases the risk of breast cancer. Is this true?

As you can imagine, this is a highly charged question. Both abortion and breast cancer are hot-button topics, and when you link them, sheesh! If you Google search this topic, you come across some very passionate folks out there. Because abortion disrupts the maturation process of the breast, it has been hypothesized that it might increase the risk of breast cancer. However, multiple studies do not support an association between abortion and breast cancer.[6] The American Cancer Society's Web site says: "The scientific evidence does not support the notion that abortion of any kind raises the risk of breast cancer." The American College of Obstetricians and Gynecologists issued a committee opinion saying essentially the same thing. While certain political factions would prefer to scare you into never getting an abortion by raising the specter of breast cancer, their allegations have no basis in science and are merely that—scare tactics.

I'm using birth control, and I don't plan to get pregnant. But I'm terrified that I will get pregnant and have to abort the baby. Can you tell me what I might expect if I wind up with a pregnancy I can't keep?

Obviously, it's always better when you can plan a pregnancy, make sure your body is in optimal condition, choose the perfect partner, and time it brilliantly so that pregnancy and parenting fits into your life. But alas, life doesn't always work this way.

Kudos to you for asking *before* you find yourself with an unwanted pregnancy. Since you're thinking ahead, please don't forget that the morning after pill (Plan B) exists, and you can purchase it over-the-counter in some states or by prescription in others. It can be a godsend for those broken condoms, skipped pills, or lusty, unplanned sexual encounters that happen when a passionate moment strikes. The sooner, the better, but Plan B is effective up to seventy-two hours after the accident.

Assuming birth control efforts fail and you find yourself faced with a painful choice, remember that it's your body, your life, your pregnancy. 'Nuff said. Of course, there's always adoption (and since my sister is adopted, I'm a big fan). But should you choose to abort a pregnancy, tell someone you love, make sure you have good support, and consider seeing a counselor. It's the hardest thing some women ever do.

If you choose early enough, you'll have the option of either surgical or medical abortion. What can you expect from a surgical abortion? The procedure, when done in the first trimester, is the exact same procedure we do when a fetus dies, but the baby doesn't pass. Basically, the doctor dilates open the cervix with blunt-tipped dilators, inserts a suction catheter that's attached to a sucking machine, and vacuums out the uterus. The whole thing takes about five minutes at the most, if all goes smoothly. Not that it isn't totally traumatic emotionally (and for most women and even the doctors performing these procedures it is). But the procedure itself isn't that big a deal.

What about medical abortion? For this procedure, medications like mifepristone (RU-486) and misoprostol are given to help you pass the baby on your own. After the medications are administered, you basically miscarry the pregnancy, with cramps, bleeding, and passage of fetal tissue.

Either way, it can be painful, emotionally traumatic, and lead to regret down the road. So please, take care of yourself. Protect yourself with birth control. Have Plan B in the bathroom, just in case. But if you find yourself choosing to have an abortion, find a skilled, compassionate doctor, forgive yourself, surrender to the choice, release your fear, and let it go.

I'm forty-five years old and have been using an IUD for decades. I'm tempted to take it out, but I definitely don't want to get pregnant. When can I safely stop using contraception?

Technically, the only way you can guarantee that you won't get pregnant is to abstain from sex or know, with certainty, that you're in menopause (which can be tricky to determine). While fertility rates at your age are very low, there's no absolute guarantee that you wouldn't conceive if you stopped using birth control. I've seen it happen. Unless you're willing to take the small but real risk of winding up with an unexpected pregnancy, you're better off keeping your IUD—or using some other form of birth control—until it's clear you're in menopause.

How late is too late to get pregnant?

When it comes to this stuff, I've learned to keep a very open mind. As we age, fertility wanes. Because there's no way to really know whether you're fertile until you try, there's no easy way to predict when it might be too late. So how late is too late to get pregnant? Well, it depends. Technically, it's too late to conceive spontaneously once you hit menopause. But fertility usually drops off long before menopause. By the age of forty-five, 87 to 99 percent of women are infertile.[7] After that, it may still be possible to conceive, but most likely you will need donor eggs implanted via in vitro fertilization (IVF).

One notable exception to the rule that you can't conceive after menopause comes to mind. My patient Maddy, who was diagnosed with premature ovarian failure (early menopause) in her twenties, was told she could never get pregnant. She never menstruated, and all of her blood tests indicated that she was in menopause. She was given birth control pills as hormone replacement, but, because she wasn't using them as birth control, she was careless about taking them every day.

One day, Maddy, who is arguably the buffest woman I have ever met, came to see me complaining that her waistline was widening and her pants didn't quite fit. She had asked her husband, who happens to be an OB/GYN, whether she looked fat. Her husband (wise man) quickly answered, "Of course not, dear."

Maddy showed up at my office, assuming she had a fibroid tumor in her uterus.

Long story short, we scanned Maddy's uterus and found a fetus growing inside. I said, "Congratulations, Maddy! Your fibroid is a boy!"

Maddy's story is one of many, but don't let that lull you into thinking you can just sit around on your laurels if you're hoping to become a mother. I absolutely believe that we make better parents by waiting until we're emotionally ready to be parents. But I hate to break it to you—biologically speaking, the sooner the better.

I don't mean to be a buzzkill to all you women out there focusing on your careers, waiting for the right partner, or sitting on the fertility fence like I did. But I think it's empowering to be informed so you can choose the life you want to live, rather than seeing yourself as a victim of fate or circumstance. It's *your* life. If you are one of those women who *must* be a mother of a child created from your own genes, don't wait. Recognize the importance of your dream and make it a priority. Remember, there's no perfect time to take a leap of faith. That's why they call it a leap of *faith*. Believe.

Is it possible to freeze your eggs if you're getting older and you're not ready to get pregnant yet?

Wouldn't that be perfect if we could harvest our eggs when we're Fertile Myrtle twenty-year-olds? Then, when we're settled into our careers, we can thaw them and have babies. Ah...that would solve everything, right?

Well, not exactly. Yes, it's possible to freeze your eggs. But lest you think it's a surefire way to guarantee your future fertility, it's important to know that it's still considered experimental.

The process is called "oocyte cryopreservation," and it requires getting hormone injections to stimulate the production of multiple eggs and going through an egg retrieval procedure to harvest the eggs (sucking the eggs out of the ovary through a long, skinny needle inserted into the vagina under ultrasound guidance). Essentially, you're making it possible to serve as your own egg donor when you finally get around to reproducing.

So why isn't every single thirtysomething freezing her eggs? First, because having eggs on hand doesn't guarantee that they will result in a baby, according to James Grifo, M.D., Ph.D., who is the program director at the NYU Fertility Center. NYU was recently able to achieve pregnancy rates equivalent to a fresh IVF cycle. In their program, they estimate that a woman who freezes her eggs under thirty-six years of age will have approximately a 50 percent chance of giving birth to a baby from a single cycle of egg freezing. As women get older, the success rates drop. For instance, a thirty-eight-year-old can expect a 39 percent rate, a forty-year-old 27 percent, and a forty-two-year-old 14 percent. The younger you are, the better.[8]

While this is good news for women with cancer who might lose their fertility—as well as women wishing to delay motherhood—Dr. Grifo acknowledges that you wouldn't want to put all your eggs in one basket, so to speak. If you freeze your eggs in your early thirties, thinking you'll be able to become pregnant later, you might wind up very disappointed if you're in the 50 percent who are unable to conceive ten years later. And keep in mind that many fertility centers cannot match the success rates found at NYU. Many are much lower. In addition, egg preservation will cost you a pretty penny. And buyer beware: There are many expensive commercial programs offering fertility preservation without any successful pregnancies. Ask how many live babies have resulted from their oocyte cryopreservation program. If they can't give you a straight answer, go elsewhere.

I've heard of some really old women getting pregnant. How in the world does that happen?

I have to admit that it concerned me when reports leaked out in the national news about a sixty-six-year-old woman giving birth. First, I worry for the health of the woman. Our bodies simply are not made to reproduce at this age. During pregnancy, our bodies must cope with insulin resistance, which can lead to diabetes; increased blood volume, which can stress the heart; weight gain, which can damage joints; and, in addition to a whole host of other stressors, there's *childbirth,* arguably a bigger workout than any marathon. To expect a sixty-six-year-old body to endure this is asking a lot, no matter how strong and healthy the woman might be. And, of course, then there's the child....

I also worry that younger women might misinterpret this hot-ticket news, thinking that they can delay childbearing indefinitely. If movie stars can get pregnant when they're fifty, why can't everyone? What many don't realize is that most of the older women who conceive do so via egg donor IVF. By combining the eggs of a fertile young donor with sperm in a Petri dish, an embryo is created. This embryo is then implanted inside the older woman, who, if all goes as planned, can carry it to term and give birth. But it's important to understand that although she carries this child, feels movement, and by all other standards becomes the mother of this child, this baby is not genetically hers.

I certainly don't judge older women who choose to go to extreme measures to become mothers. In fact, I've delivered dozens of them. The way I see it, if a woman understands the risks and chooses to assume those risks, it's her life—her choice—and I support her. But it's important to understand that these pregnancies almost never happen naturally. And there's no guarantee that it will be successful.

My friend Ellie, who is forty-four and has been unable to conceive, spent seventy thousand dollars to hire a young, healthy egg donor and undergo egg donor IVF in hopes of fulfilling her dream of becoming a mother. Her donor turned out to be a bit of a dud, and the only three eggs that were successfully fertilized were implanted into Ellie's uterus. Two weeks later, she had a period and watched all of her hopes, dreams, and hard-saved money flush down the toilet with her menstrual blood. The success stories make the news, but stories like Ellie's are more common.

If you're still on the fence about whether you want to become a mother, I challenge you to ask yourself this question: Must you become a mother by reproducing your own DNA? Search your heart. Listen to the voice of inner knowing within you. If the inner voice is yelling, *Yes!*, don't delay, and go to extreme measures if you must. But if you're open to other ways of becoming a mother, take your time. As the sister of an adopted child, I can tell you that a mother who adopts is every bit a mother and my sister has never felt like anything but a sister to me. Trust me, there's abundant love to give and receive, no matter how you become a parent.

How can sperm make me pregnant if it all drips out of me right after sex?

Trust me. Those little suckers can swim! With little tails that help their locomotion (humming the old song in my head right now, "C'mon, baby, do the loco-motion"), sperm can hightail it right where they need to go. While some abnormal sperm with two heads that spin in circles may leak out, what drips all over you consists mostly of the other components of semen—primarily proteins, citric acid, minerals, enzymes, sugars, and other components necessary to aid sperm in their mission. These components of semen help nourish and support the sperm, but once they've served their purpose, they drip back out, resulting in the proverbial wet spot.

How fast do sperm swim?

About 5mm per minute, which means they can travel about 5 sperm body lengths per second. (Damn, that's fast!) To give you some idea how fast that is, if sperm were salmon, they would be swimming about 500 miles per hour. If they were whales, make that 15,000 miles per hour. But remember how tiny they are. It's not like they're running some sperm marathon inside you. Because of their itty-bitty size, it can take a day or two to swim all the way up to the egg.

How do they know which way to go? They follow heat like little heat-seeking missiles. The farther in they go, the warmer the body gets. The egg also does a little mating dance for the sperm, emitting substances that attract sperm, all in the hope that one lucky sperm will pierce through the egg's protective shell and be the lucky guy who fertilizes the egg.

My friend told me I should hold my legs in the air for fifteen minutes after sex if I want to get pregnant. Is this true?

While it won't necessarily help you conceive, you might wind up with some rock-hard abs before pregnancy stretches them out! There's no solid evidence to support the myth that certain positions during or after sex will improve fertility.[9] But fertilization rates may be higher if sperm hang out in the vagina for ten to fifteen minutes.[10] And certainly, there's no harm in it. It makes logical sense to choose sexual positions that optimize the amount of time the sperm are near the cervix. So gravity-defying sexual positions, such as standing, woman-on-top, or swinging from a trapeze, may be less than ideal. Many fertility doctors recommend the tried-and-true missionary position. Do what feels right to you.

*My husband and I are trying to get pregnant,
but he's so freaked out, he can't get it up. What
should I do?*

My husband wigged out, too. Right around that time of the
month, I grabbed his hand, yanked him into bed, and squealed,
"I might be ovulating! Give it to me, big boy!" Normally up for a
good romp any time of day, poor Matt's erection deflated like a
very wilted carrot.

When I looked disappointed, Matt said, "I've spent my
entire adult life trying to avoid getting women pregnant. Can
you just give me two seconds to get used to the idea?" Okay, so
maybe I was a bit overly enthusiastic about the whole ovulation
thing.

It made me realize that while we women may be receptacles
for baby-making, guys have to actually perform. Assuming that
you've laid everything on the table and your partner actually
wants to make a baby with you, I'd recommend keeping the
whole ovulation thing on the down low. Just because your LH is
spiking doesn't mean you need to announce it. Seduce him often
when you're trying to conceive, but leave him guessing about
which time might be *the* time.

You might also get creative about ways to help him get it up.
Pull out some porn. Use your feminine wiles. Get down and
dirty—whatever it takes. And keep it light. Just because you're
having sex for procreation doesn't mean you can't still use it for
recreation.

If we're trying to get pregnant, do we have to have sex every day?

While some couples view baby-making as a great excuse for bopping like bunnies, others quickly find that sex becomes a grind. Before you know it, he's reading the sports section and you're doing your nails while you're having sex. How can you find balance? What's the ideal sexual frequency for making a baby?

Your best bet is to have sex one to two days before ovulation. Couples seem most likely to get pregnant when they have sex approximately forty-eight hours around the window of ovulation. More is not necessarily better. Data suggests that semen quality is at its peak if you hold off for two or three days between ejaculations. (Guys, that means no jerking off in between. Sorry!) Consider your menses a break during which you only have sex if you feel like it, but start having regular intercourse when your menses ends, at least two or three times per week. And make sure you're having sex every forty-eight hours around midcycle. Then you can settle back down after the window of ovulation passes.

Are there sex acts we should avoid if we're trying to make a baby? Oral sex? Vibrators? Lubricant? Whipped cream?

When you're doing the deed for procreation purposes, sex may start to feel a bit dull. So how can you spice up your sex life without reducing your fertility? Experimenting with sexual positions can be fun, and no one position seems to be significantly better for fertility, so go nuts with the Kama Sutra! Or try out role-playing, sexy seductions, sensual massage, or erotic films, magazines, and literature. Thankfully, vibrators are on the menu. As long as you're being gentle, there's no reason not to use a vibrator when you're trying to conceive.

As for whipped cream and other fun foodstuffs, you might want to limit them to breasts and belly buttons, since some foods may impair sperm motility if used on the genitals. Also, some lubricants, such as Astroglide, K-Y Jelly, Touch, olive oil, and (sigh) saliva seem to impair sperm motility in a Petri dish. Although there's no clear evidence that they impair fertility in real life, you might want to avoid lubricants if you're trying to be über-cautious.

All that said, don't forget that you've got to live a little. Unless you're fertilely challenged and pulling out all the stops, I say don't worry too much about the dos and don'ts of trying to conceive. Sure, it makes sense to do your homework and make modifications to optimize your fertility. But don't forget to have fun, let your freak flag fly, and let go of the rest.

If I'm trying to conceive, do I have to stop drinking alcohol while we're trying, or is it okay to indulge until after I get a positive pregnancy test?

Of course, I'm a doctor, so I'm supposed to tell you to abstain from alcohol altogether. That's certainly the safest option. But in addition to being a doctor, I'm also a woman who understands that you're already going to give up alcohol for nine months and you may not wish to abstain for the entire length of how long it takes you to conceive.

So how can you toast to life while making sure your body is the optimal vessel for the child you're inviting in? If you wish to indulge, do so during the first part of your cycle, starting from the first day you bleed (day 1). You're likely not pregnant during this time, so if you're planning to imbibe, this is the best time. After ovulation, usually around day 14 of your cycle, there's a chance you could be pregnant. Because fragile fetal organs develop the first few weeks of pregnancy, it's best not to expose those delicate cells to the effects of alcohol. If you do choose to indulge, be sparing during those last two weeks of your menstrual cycle, when you might be pregnant without knowing it yet. Sip a glass of wine, but if someone's handing out shots, hand yours off to someone else. If you happen to drink during the second half of the cycle, not knowing you are pregnant, don't beat yourself up about it. It happens all the time when people wind up with unplanned pregnancies. Chances are that everything will be fine.

If I'm trying to get pregnant and my period was a few days late, does that mean I might have gotten pregnant and miscarried?

Yes. It's possible. Many women not using birth control get pregnant and never even know it. You may conceive, but your period isn't late and you bleed right on schedule. Maybe you're a day or two late and then your period is slightly heavier than usual. These kinds of pregnancies happen all the time.

One study measured daily pregnancy tests in sexually active women not using birth control. Doing so, they were able to pick up very early pregnancies and found that many never come to fruition. Among the positive pregnancy tests they found, 31 percent of those pregnancies resulted in miscarriage. Of those that miscarried, 70 percent happened before the woman ever would have known she was pregnant.[11] If you include eggs that get fertilized but never implant, about half of pregnancies miscarry.[12] Most of these miscarriages, however, are never clinically diagnosed.

So yes, if your period was a few days late, you might have been pregnant and lost the baby. But don't forget that miscarriages are usually nature's way of making sure the pregnancy is healthy. Usually, there's nothing you did to cause it and nothing you can do to prevent it. Even if you did miscarry, there's no reason to believe the next one won't be the one that sticks.

Are home pregnancy tests as accurate as the ones you get at the doctor's office?

Maybe, maybe not. It depends what kind of test your doctor is using. But the real difference is that those who work at medical offices are experts at reading these tests, whereas most people reading home pregnancy tests are not. Sure, peeing on a stick doesn't sound like rocket science. But studies demonstrate that home pregnancy test failures are most often a result of people reading them wrong. And most errors occur from reading a positive test as negative, not the other way around. So if you read your home pregnancy test as positive, chances are that you're pregnant. To quote the movie *Juno*, when the store clerk sees

Juno shaking her positive home pregnancy test, "This ain't no Etch A Sketch. This is one doodle that can't be undid, Homeskillet."

How early are home pregnancy tests positive?

I know how tempting it is to pee on that stick the day after you ovulate, but do yourself a favor and wait. You'll only waste pregnancy tests and increase your anxiety. Using an extremely sensitive urine assay, 90 percent of pregnancies are detected on the day you would expect your next period (the first day of a missed period), 79 percent are detected two days before, 97 percent one week later, and 100 percent eleven days later.[13] Home pregnancy tests may take longer to register as positive, especially to the untrained eye. If you really want to be certain, wait at least a week after your missed period before testing. There's no harm in delaying the diagnosis a few days and it's more likely to be accurate. If you're dying to know, wait at least two weeks after the time you would have ovulated (which should be right around the time you would expect your next period). If it's negative and you're suspicious that you're pregnant, repeat a pregnancy test a week later, just to be sure.

CHAPTER 9

Pregnancy

WHEN I SOLICITED QUESTIONS FOR this chapter, I was blown away by the fact that most of the pregnancy-related questions were about what we *can't* do when we're expecting. Don't get in a hot tub/drink caffeine/eat deli meat/have a glass of wine/get a massage in the first trimester/go running/eat tuna. The minute that pregnancy test turns positive (or even before), we're barraged with a whole litany of obsessive-compulsive dos and don'ts. And if you choose not to obey every single rule, you'd better prepare to get some serious stink-eye from people.

Can I just say this? Pregnancy is not a disease. Repeat it out loud. *Pregnancy is not a disease*. It's a natural state of being for women, and while I certainly don't condone being reckless, I think we've gone way overboard in how we approach being pregnant. It's no wonder we get so neurotic.

Don't get me wrong. It's not like I was blissfully relaxed and mentally healthy during my pregnancy. While being an OB/GYN certainly arms you with a load of knowledge, it's possible to know too much. I was keenly aware of every single thing that could possibly go wrong, but I tried to stay focused on the ultimate goal—healthy mom, healthy baby.

I treated my pregnancy like a Ph.D. dissertation. Each week of the pregnancy was a chapter to be completed; each trimester, a section to hand in to my advisor. Every lab test was an A to earn. Each ultrasound was an affirmation that the thesis was coming together nicely—each sentence constructed carefully, every chapter researched and footnoted.

To help allay my growing anxiety, I started jamming the ultrasound probe up my vagina several times a week, starting the minute my pregnancy test showed the little pink plus sign. There were no locks on the exam room doors in my office, so I had to push my naked bum against the door to keep it closed, while scanning myself with the ultrasound machine. Finally, I saw what I was longing to see. When I hit the five-week mark, there it was: a perfect little gestational sac on the ultrasound monitor. A week later, I saw the heartbeat flickering away. I mentally marked off "Living fetus. Check."

At twelve weeks, my doctor measured the nuchal translucency—the thickness of the fetus's neck—which was normal. Check. A few weeks later, I had my amniocentesis. 46XX. Normal amnio. Girl baby. Check.

I started paying attention to see if I could feel my baby move. In spite of a decade of experience in obstetrics, it was the one thing I could never imagine. Patients told me, "It's like butterflies. Like a fluttering."

But they were *way* off, at least for me. I first felt painful little twinges, right inside my cervix. When I grabbed the ultrasound and stuck it on my belly, I discovered that, sure enough, baby Siena was tap-dancing on my cervix. Every time her foot hit my cervix, I felt the twinge. Butterflies, my ass. Anyway, I checked that off my list as a reassuring development. Fetal movement. Check.

At twenty-four weeks, another checkmark. My pregnancy had survived to fetal viability, the gestational age when fetuses have a chance of surviving outside the mother. Then my gestational diabetes test came back normal. Check. The only mar on my pregnancy report card was the placenta previa that hovered over my cervix, threatening to prevent me from having a vaginal delivery. I have to admit that I felt a tad relieved. I could just schedule my C-section into my busy life. Check.

Of course, you know what they say about the best-laid plans....

When I was twenty-eight weeks pregnant, my father was diagnosed with brain cancer. With tumors metastasized all over his body, he would be lucky to survive three months, which, as fate would have it, meant he was likely to die on my due date. So even though my placenta previa moved out of the way and I might have delivered naturally, my husband and I opted to push forth with a scheduled C-section. Dad was right there at the hospital, waiting to hold my daughter on her birthday. Seeing my father, with his radiation-bald head, holding the tiny person I had just made, shifted something in the tectonic plates of my life. There's nothing like the juxtaposition of birth and death to remind you of the circle of life. Dad kissed my daughter good-bye two weeks later. With nothing left unsaid, he went to sleep and didn't wake up again on what would have been my due date.

My pregnancy became a lesson in life for me. When what I came to call my Perfect Storm passed, I found myself forever changed, as we all do once we usher a new life into the world. Checklists no longer serve me.

Pregnancy is really a lesson in surrender, though it tends to shake up the control freak within us all. It begins with trying to conceive. You can take your temperature, pee on ovulation kits, and do the deed on all the right days. But you can't effort yourself into becoming a mother. It continues during pregnancy, when you learn that you can make checklists, you can struggle for control, and you can jump through all the right hoops. But you can't control the outcome. If I knew then what I know now, I would go back in time and tell myself to just let go, to go with the flow, and to trust the Universe. But I suppose we all have to learn our lessons in time.

I think, on some level, we realize this, which makes us cling all the harder to the few things we might control. I hear you, sisters. In this chapter, I will try to shed some light on these issues for you. But keep in mind that the sooner we realize that we can't control every aspect of our lives, the better parents we become.

How quickly will I have symptoms of pregnancy?

Every woman and every pregnancy differs. Some of my patients insist they notice subtle pregnancy symptoms as soon as a week after conception. Others swear they had no clue they were pregnant until they suddenly wound up in labor and a baby popped out. (Go figure!) Usually, the first sign that you're pregnant is a missed period, which occurs two weeks after conception for those with regular cycles. Breast tenderness, nausea, frequent urination, and food aversions commonly appear around four to six weeks from the time of conception. If you're pregnant and not experiencing symptoms, you may be one of the fortunate few who carry a normal pregnancy to term without symptoms. Or it may indicate that your hormones aren't rising appropriately, which puts you at risk of miscarriage. The best way to find out is to talk to your doctor.

Can I dye my hair when I'm pregnant?

I once overheard a patient ask my former partner this question. My partner replied, "What are you afraid of? That your baby will come out blond?" Cracked me up. For many years, doctors have believed that dying your hair when you're pregnant is probably just fine, but like many issues relating to pregnancy, good data is limited because pregnant women are so hard to study and the data we have always seems to be changing. The Organization of Teratology Information Services (OTIS), which provides information on potential risks in pregnancy, offers some reassurance on the subject of hair treatments and pregnancy. In animal studies that used doses one hundred times higher than would normally be used in a human, hair dye didn't increase the risk of negative outcomes. Permanents and chemical hair straighteners also appear to be safe. Even part-time cosmetologists who are surrounded by these chemicals do not appear to be at significantly greater risk. (Full-time cosmetologists may be at slightly higher risk of miscarriage, but it's unclear whether this is related to hair dye or long hours of standing behind the chair in the salon.[1])

However, new information is emerging on the safety of

chemical exposures during pregnancy. According to Joanne L. Perron, M.D., FACOG, a postdoctoral fellow in the Program on Reproductive Health and the Environment at UCSF, new research suggests that exposure to chemicals found in many personal products such as hair dyes may affect the fetus at the genetic level, even at tiny doses. These changes in the genetic functioning of our babies may increase the risk, later in life, of numerous conditions such as neurodevelopmental disorders, reproductive disorders, and cancer.

So what's a woman with roots to do? Check out the Environmental Working Group's Web site (www.ewg.org) to find hair dyes that are safer. Or consider going au naturel during early pregnancy when the fetus is most susceptible to toxins. While the jury's still out on hair dye, you'll want to do what you can to protect your child's fragile DNA for the future.

Is it true that I can't get in a hot tub when I'm pregnant?

Doesn't it seem like they try to rob you of all life's simple pleasures when you're pregnant? I mean, since when did your old friend the hot tub turn into a baby killer? After all, when you're pregnant, your whole body hurts, you can barely sleep, and your life is about to get turned upside down. A little warm-water massage and stress relief might do you some good.

When I was thirty-six weeks pregnant, I took my dying father to a natural hot springs spa famous for their healing mineral water baths. We must have made an unlikely pair, me in my bikini because I hadn't bothered to buy a maternity bathing suit and Dad with his radiation-bald head. Some woman had the nerve to approach me when I was floating with Dad in the hot springs and say, "Are you trying to kill your baby in here?"

I looked her in the eye and couldn't resist saying, "I'm giving the baby up for adoption."

Okay, so it wasn't one of my stellar moments. But I thought she had some nerve trying to tell a pregnant OB/GYN with back pain and a dying father what she should and shouldn't do.

The rumor about hot tubs being the gestational devil stems from some limited data that suggests that a high fever early in pregnancy may lead to neural tube defects (spina bifida) and, possibly, miscarriage. To cause these abnormalities, evidence from animal studies indicates that the core body temperature must be sustained at 102°F or higher for an extended period of time.[2] Somehow, this data has gotten bastardized into a recommendation that all pregnant women avoid all hot water. Even my physician friend called when she was pregnant to ask, "Is it okay for the shower water to hit my belly, or should I only shower belly out?" I mean, seriously. We've gone a little overboard.

Even the highly regarded OTIS agrees that short-term hot tub use is most likely perfectly safe in pregnancy, especially after six weeks' gestation. They recommend limiting hot tub use to no more than ten minutes at a time, in order to avoid raising your core body temperature to 102°F.[3] Even then, they admit that normal hot tub use does not appear to increase pregnancy risk past 6 weeks gestation.

What do I tell my patients? I suggest turning down the temperature on the hot tub. If the hot tub is set at 101°F instead of 104°F, you should be just fine. If you want to keep it at 104°F, don't stay immersed longer than ten minutes. Get out, cool off, then get back in if you feel like it. But make sure to drink lots of water—dehydration is pregnancy's enemy and can lead to pre-term contractions. So guzzle up if you're planning to soak.

What about saunas? The same holds true. Avoid raising your core body temperature above 102°F and you should be A-Ok.

If you're pregnant and interested in hot-tubbing, be most careful until six to eight weeks' gestation, but it's likely okay to go for it after that. If you had a high fever or spent time soaking before you knew you were pregnant, chances are nothing's wrong, but you might want to talk to your doctor about screening tests for neural tube defects, such as AFP. Using a hot tub during my pregnancy helped ease aching muscles and did wonders for my broken heart. If it doesn't resonate with you and you feel like skipping it, skip it. But if the hot tub beckons, I say go for it, girlfriends. Just do so with care.

Is it true that sex can stimulate labor? If so, will having sex make me deliver early?

Ah, if only it was that easy. While we OB/GYNs like to recommend sex to pregnant women at term to try to hurry up labor, there's no real data to support its ability to stimulate labor.[4] Certainly, the theories make sense. We know that prostaglandins, such as the drug misoprostol, ripen the cervix and stimulate contractions. And semen is full of prostaglandins. Plus, orgasm stimulates release of oxytocin, which can stimulate labor. Not to mention that messing around with the cervix can trigger contractions and get things started. But so far, studies do little to support this recommendation.

But that doesn't mean we haven't all heard stories of how sex triggered labor. Australian physician Jules Black, M.D., put it best. He said sex and labor are like fruit falling off a tree. Early in pregnancy, the apple hangs tightly to the tree. Even if you shake the branch, it's unlikely to come off. As the apple ripens and becomes more succulent, it starts to loosen its hold on the branch. But it's not until it's ripe and ready to fall off that the wind blowing might knock the apple off the tree. Similarly, sex is unlikely to stimulate labor before it's meant to happen. But maybe, when the pregnancy has fully ripened, sex will push you over the edge, like the gentle breeze that causes the apple to fall.

I guess we docs figure there's no harm, and pregnant women are so desperate to do *something*. So why not? Everyone is different. It just might work for *you*. Might as well take advantage of the excuse before welcoming your little one into the world.

I've heard that you can induce labor just from playing with your nipples. Is this true?

Yup. Sure is. In fact, most hospitals have official nipple stimulation policies intended to help induce contractions. Can't you just imagine Nurse Ratched barging into a labor and delivery suite in her starched white suit, stripping off your bra, barking orders, and going to town on your poor brown nipples? Swear to God, I've seen it happen. Although most of the nurses I've worked with take a hands-off approach, preferring to teach a woman how to do it herself under medical supervision.

But don't try this at home. Nipple stimulation—at least the way it is done in the hospital—aggressively pulls and tugs at the nipples, which serves to release natural oxytocin, a labor stimulant. In a woman whose cervix is favorable (soft, beginning to dilate, and shortened), nipple stimulation can be an effective way to induce labor.[5] But some pregnant women are very sensitive to nipple stimulation, and it's possible for the uterus to contract *too* much, leading to fetal distress. While there's no reason to avoid sexual nipple foreplay during a healthy pregnancy, you don't want to try to do Nurse Ratched's job unless you're being monitored by a health-care professional.

I went to the spa to get a prenatal massage when I was ten weeks pregnant, and they refused to give me one. Why?

There's absolutely no evidence that massage leads to miscarriage. Then why are spas so fussy about first-trimester pregnancy massage? Because their lawyers have freaked them out. Miscarriage is common in the first trimester. Depending on how you define it, the miscarriage rate ranges from 8 to 31 percent.[6] If you happen to get a massage and then you miscarry, you might blame your massage and sue the spa.

Don't get me up on my soapbox, but come on, people! If you get a massage and then you miscarry, it's not because of your massage. Miscarriages just happen, and chances are that your miscarriage is completely unrelated to anything you did. It's more likely that nature is protecting you from carrying an abnormal baby to term.

But those who are quick to sue have ruined it for the rest of us. So if you're in your first trimester and you want to get a massage, go for it. The only massage technique I might avoid is Maya abdominal massage, which manipulates the abdominal and pelvic organs.

Can the baby feel us if we have sex while I'm pregnant?

Uh, maybe. But the more important question would be, "Does the baby care?" If the baby is still a wee fetus, it probably doesn't even notice when the uterus rocks around a bit. After all, the same thing happens when you're running (bounce, bounce, bounce). As the fetus gets older, tickling its head during a pelvic exam produces reassuring signs in the fetus's heart rate. The heart rate rises, resulting in an "acceleration"—a sign of oxygenation that signals fetal well-being. So you might say that the baby likes having its head tickled.

Some couples worry that they will cause psychological damage by having sex during pregnancy. Actually, the fetal mind has no clue what's going on down there. For all the fetus knows, it's being rocked gently to sleep. So rest assured. Having sex during pregnancy will not mess up your baby's psyche. If anything, it will bond you to your partner, helping you endure the rough months of physical recovery, sleep deprivation, and life change that may lie ahead.

Must I give up my morning grande cappuccino when I'm pregnant? How great is the risk if I can't give it up?

While cutting back is prudent, you don't have to quit cold turkey. One study showed a higher risk of miscarriage in women who drink more than one hundred milligrams of caffeine per day. The higher the dose, the greater the risk, with most studies suggesting the risk is most evident at levels higher than three hundred milligrams per day.[7] Based on their review of the literature, the Organization of Teratology Information Services (OTIS) recommends that pregnant women limit caffeine consumption to no more than two to three hundred milligrams per day.[8] Most studies suggest that levels lower than one hundred milligrams do not increase the miscarriage rate.

How does this translate? Well, a Tall (twelve ounce) Starbucks latte contains about seventy-five milligrams of caffeine. For a grande latte (sixteen ounces), double that. A twelve ounce Coke contains forty-six milligrams. Most studies suggest that lower levels of caffeine do not increase the miscarriage rate.

And what about birth defects? Can caffeine make your baby have two heads or grow extra fingers? No. Even large amounts of caffeine have not been shown to increase the risk of birth defects. But if you're a caffeine addict, your baby may be born with a faster heart rate, tremors, and insomnia. Do you really want your baby's life to start out that way? And keep in mind that caffeine is a potent diuretic that will make you pee even more like a wild monkey than you already are, which can cause dehydration and stimulate preterm contractions. All in all, it's just not a good idea.

The moral of this story is that your pregnant body is simply better off without caffeine, if you can avoid it. But if you're like I was during my pregnancy and cutting out caffeine is out of the question, try to wean yourself down to one cup o' joe per day, at least until you make it past the first trimester.

Is it really bad if I drink a glass of wine here and there when I'm pregnant? I'm not talking about getting wasted, but if I feel like sipping a little chardonnay from time to time, am I a terrible mother?

Some of you are old enough to remember how much times have changed regarding this issue. Back in the fifties, pregnant women sat around with beehives, poodle skirts, cigarettes, and martinis. And why wouldn't they? For decades, doctors prescribed alcohol for therapeutic reasons, to treat preterm labor and to relax the uterus. Dialogue about the dangers of alcohol in pregnancy hadn't even begun.

In the 1960s–70s, IV alcohol was regularly used in pregnant women to stop contractions. These women would get so blitzed, bucking and puking, that they would have to be restrained in their hospital beds. Not until reports of fetal alcohol syndrome began to surface in the late 1970s did alcohol begin to be regarded as potentially damaging to babies. In 1981, the surgeon general put out the now-ubiquitous warning about alcohol and pregnancy. Suddenly the rhetoric in the medical literature shifted from talking about alcoholic mothers and fetal alcohol syndrome to a generalization that even small amounts of alcohol in pregnancy might be toxic.[9] Ever since, pregnant women, understandably, have been confused.

Sure, it's easiest to recommend abstinence during pregnancy. If you don't consume any alcohol, you can guarantee that alcohol will not influence your baby. But is abstinence absolutely necessary? What about a glass of wine from time to time? Good question.

We know for sure that regular, high-dose alcohol consumption can cause *fetal alcohol syndrome*, a severe disability characterized by mental retardation, birth defects, developmental delay, and a whole host of other abnormalities. What's far less clear is how much alcohol may cause what we call "fetal alcohol spectrum disorders"—which, in addition to the abnormalities associated with fetal alcohol syndrome, include more subtle problems, such as learning disabilities, behavioral issues like attention deficit hyperactivity disorder (ADHD), and psychiatric conditions.

One study did tread lightly into the whole issue, assessing whether the children born to women who drank up to one or two drinks per week during their pregnancies were at higher risk of behavioral difficulties or cognitive deficits. It found that these children assessed at three years of age in this study were not more likely to have behavioral difficulties or cognitive deficits when compared to the children of mothers who abstained.[10] But our data still falls short in providing guidelines about safe alcohol limits.

What I can tell you is that many OB/GYNs I know admit to giving their pregnant patients the go-ahead to enjoy a drink on occasion. Many of these same docs imbibe alcohol themselves when they are pregnant. But this kind of permission happens in whispers, behind closed doors. I can't find many docs willing to take a stand publicly. You can understand why. Without good data to support us, it's hard to mess around with something as fragile as a new life. It's easier—and probably safer—to err on the side of caution.

So I apologize, but I'm afraid I have to conclude that if you have a glass of chardonnay here and there, you do so at your own risk. It's probably just fine to enjoy the occasional glass of wine. There's a big difference between being reckless (a definite no-no) and having one drink at dinner. Millions of women throughout time have done so and then given birth to healthy, happy children (myself included). But there's just no data to help us make safe recommendations.

Here's how I counsel my patients: Are you one of those women who freak out and will blame yourself if anything ever goes wrong with your child? Will you beat yourself over the head if you drink a glass of wine in pregnancy and then your kid is diagnosed with autism or has a learning disability? Chances are, if this happens, it will be unrelated to the alcohol you consumed. But will you wallow in guilt, asking impossible "what if" questions? If you're this kind of person, skip the alcohol altogether. It's just not worth it. Why torture yourself? If that doesn't sound like you, go ahead. Make your own choice.

But trust me on this one. If you do decide to throw caution to the wind and toast with a glass of champagne, do it at home or in France. If you live in the United States and you're out in public, you're likely to get some sideways glances if you're obviously pregnant and drinking. Good or bad, that's the world we live in.

What foods should a pregnant woman really avoid? Brie? Sushi? Swordfish? Lunch meat?

With morning sickness, heartburn, and a stomach that gets pushed up into your diaphragm, you may find it hard to eat *anything* when you're pregnant. Add to that the laundry list of dietary no-nos you'll read about in most books, and it's a wonder that pregnant women eat at all. So what *can* a pregnant woman eat?

I'm a big fan of moderation. Everything in moderation— even moderation. Pregnancy is not a good time to eat the same thing over and over again. A balanced, varied diet of whole foods, heavy on the veggies, fruits, lean proteins, and whole grains, will serve you well. Unfortunately, even if you eat all organic, hormone-free, wild-caught, free-range food, you can't eliminate all risk. You also can't live in fear.

The main reason you hear so much fuss about diet in pregnancy is because the immune system of a pregnant woman is naturally suppressed. It's the body's way of keeping you from fighting your baby as a foreign body. This natural immunosuppression makes you more susceptible to infections, and when pregnant women catch certain infections the baby may be at risk. Also, because the fetus is so fragile, certain exposures, such as mercury, may affect their developing bodies.

To limit the risk to you and your baby:

1. Stick to fully cooked meats, fish, poultry, and eggs. Which means you can eat sushi, just pick the cooked shrimp, crab, or eel rolls.

2. Avoid unpasteurized dairy products, such as fresh cheese you might find at a local farmer's market. If you can buy it at a regular grocery store, it's probably pasteurized. But some specialty cheeses, especially foreign imports, may not be. If you're not sure, go to a good cheese shop and ask.

3. Be mindful of the fish you choose to eat. The U.S. Food and Drug Administration (FDA) advises pregnant women to avoid the fish with the potential for high mercury levels (shark, swordfish, tilefish, king mackerel) and limit fish with lower risk to two meals/week.

4. Wash all fruits and veggies thoroughly, and try to eat organic if you can. If you don't have access to organic produce or can't afford it, avoid the "dirty dozen"— peaches, apples, bell peppers, celery, nectarines, strawberries, cherries, kale, lettuce, grapes, carrots, and pears, which may be most heavily laden with pesticides.

Many of the dietary "nos" you read about relate to the bacteria *Listeria monocytogenes.* Unpasteurized cheese, deli meats, pâté, unpasteurized milk, and smoked seafood have all been linked to listeria, which can cause an exceedingly rare infection that may lead to fetal death or neonatal infection. Toxoplasmosis is another concern, and may be contracted by eating soil-contaminated fruits and veggies or raw, undercooked, or cured meat. If you choose to eat these foods, your risk of having problems with your pregnancy is still very low, but if you want to cover all your bases, you might steer clear.

What causes morning sickness? And why do they call it morning sickness when it lasts all day long?

I hear you. You're yakking all day long and your coworker says, "Got a little morning sickness?" Green-faced and sour-mouthed, you want to smack her upside the head and remind her that it's 4 P.M. and you've been throwing up everything you ate all frickin' day long. Trust me, I've seen it over and over.

Nausea is so common during pregnancy that if a patient of reproductive age shows up at my office complaining of nausea, the first thing I do is perform a pregnancy test. (I got burned too many times. You look awfully silly if you start heading down the path of diagnosing some gastrointestinal disorder when the woman is merely pregnant.) Nausea, with or without vomiting, occurs in 50 to 90 percent of all pregnancies.

As for the term *morning sickness,* it's a total misnomer. I swear, we docs didn't make it up, and this lay term does nothing to accurately describe pregnancy-related nausea. While many people notice nausea more severely in the morning, 80 percent of those who describe nausea during pregnancy report that it happens all day long.[11]

As for what causes it, nobody quite knows. The leading theory is that pregnancy-related nausea has something to do with rising levels of a hormone in the body called HCG. Whatever the cause, it's actually a healthy sign during the first trimester. While lack of nausea shouldn't concern you, the presence of nausea bodes well. Those with nausea are less likely to miscarry or have a stillborn baby.[12] So look at the bright side. While you may curse it for impacting your quality of life, try to reframe your thinking. Nausea in pregnancy is just nature's way of signaling to you that your baby is healthy and growing.

Do I have to get rid of my cat now that I'm pregnant? If not, is it okay if the cat jumps on my belly?

As the ex-wife of a veterinarian, let me say this: *Please* do not get rid of your cat because you're pregnant. Vets get asked to put healthy animals to sleep when a woman conceives, which is ludicrous. Spare poor Pussy, who has every right to continue living happily ever after as a part of your family.

The hoopla stems from the fear of *toxoplasmosis,* caused by a ubiquitous parasite, *Toxoplasma gondii,* which can cause fetal infection if a woman acquires this infection for the first time when she's pregnant. Up to 50 percent of us have already been exposed in the past, so this is not a rare phenomenon.[13]

Where does Pussy fit in? If Pussy gets toxoplasmosis for the first time, she will shed the oocysts of the parasite in her poo for a period of one to three weeks. These oocysts become infectious one to five days later. If the litter box sits around long enough to become infectious and you get Pussy's poo on your hands and then ingest these oocysts, you may acquire toxoplasmosis. If you've never had it before, your infection can cause serious brain and eye abnormalities in the baby.

The simple answer is don't put Pussy's poo in your mouth. The best way to avoid this is to have someone else change the litter box. Changing the litter box daily also reduces the risk. (Remember, it takes a while for the oocysts to become infectious, so the more you change the litter box, the lower the risk.) If you must change the litter box yourself, wear gloves and wash your hands carefully afterward. And if Pussy is an indoor/outdoor cat, you might consider keeping her inside to reduce the chance that she will pick up the parasite. But I beg you (and vets everywhere agree!), don't give up on poor Pussy.

As for jumping on your belly, unless Pussy is a cheetah, you should be okay. Big dogs and children should avoid causing direct trauma to the belly. You certainly don't want your pregnant belly turning into a pet or toddler trampoline. But gentle romping should be just fine.

Since I got pregnant, I'm as horny as a teenage boy. What's the deal?

I know what you mean. When I was pregnant, I constantly wanted to jump my husband's bones, even though I was self-conscious about my changing body. If we could bottle up that pregnancy libido and sell it as an aphrodisiac, we'd make millions!

Why is this the case? When you're pregnant, levels of estrogen and progesterone go through the roof, which causes increased blood flow to the pelvic area, greater sensitivity of the breasts, and more vaginal lubrication. Bingo. All rolled together, it acts like a little natural Viagra. In addition to higher libido, some women also report more intense orgasms during pregnancy, most likely as the result of increased blood flow. And of course, great orgasms are likely to make you want more sex.

I also found that pregnancy made me more aware of my body. Usually, I have a tendency to live in my head. My husband and I are having sex and I'm thinking about tomorrow's to-do list. But when I was pregnant, the whole experience was so somatic—you become so aware of every little bodily function because everything changes. I found it much easier to be physically present within my body, which can do wonders for your sex life.

If pregnancy has the opposite effect on you, you're not alone. Some women feel self-conscious about their evolving shape, which leads to decreased libido. And some are just too downright uncomfortable to even think about sex. Either way, it's normal. Your whole life is changing. It's no wonder your sex life changes, too.

Is there any way to prevent stretch marks?

It's bad enough that our babies forever alter our figures. But must we endure garish stretch marks to boot? While there's no guaranteed way to avoid stretch marks during pregnancy, avoiding rapid weight gain is the ticket. If you gain eighty pounds during your pregnancy, you're almost guaranteed to get stretch marks. If you slowly gain the twenty-five to thirty pounds recommended during pregnancy, stretch marks are less likely (but still possible). If you do wind up with stretch marks, blame your parents. Whether or not you get stretch marks seems to be largely genetic.

As for prevention, the beauty industry leads you to believe you can prevent stretch marks if only you apply the right potion. In spite of the rumors, cocoa butter does not prevent stretch marks.[14] The only cream I could find that has evidence to prove that it actually helps contains centella asiatica extract, alpha tocopherol (vitamin E), and collagen-elastin hydrolysates.[15] But when I was pregnant, I couldn't exactly find this at my local Walgreens.

The good news is that stretch marks shrink and fade with time. If you're self-conscious about your stretch marks, tretinoin (Retin-A) may be used postpartum to help reduce the appearance of stretch marks. Laser treatment can also help.

Personally, I recommend acceptance. So your body changes after having a baby—so what? It's a given. Remember that true beauty lies within and can't be touched with stretch marks. It's worth the sacrifice to bring forth a new life into this world.

*I've been sleeping with two different guys
and now I'm pregnant. How can I find out
who the father is?*

Oh, boy. Have I heard that question before. You'd be surprised how many of my patients aren't sure who the daddy is.

I'll never forget Brooke, whom I met when she showed up pregnant. She sat hand in hand with her husband, Bruce, and when I pulled out my little pink wheel to give her a due date, they both leaned *way* forward.

I said, "Looks like October 23, plus or minus a few days." The date confirmed what I had seen on a vaginal ultrasound.

Brooke and Bruce glanced at each other. "What do you mean, 'plus or minus a few days'?" Brooke flipped furiously through her calendar.

I explained that you can never know the exact date of conception, unless you've gotten pregnant with IVF. Brooke handed me her calendar and said, "Can you tell me exactly which dates I might have conceived?" I did the math and pointed out a window of dates on her calendar. "Most likely, it was sometime between here and here."

She pointed to a few other dates on her calendar. "No chance it could have been here? Or here?"

I shook my head. "Not likely."

Bruce turned green and Brooke burst into tears. I wondered what I was missing. After a few minutes of awkward silence, Bruce said, "We should tell her."

Brooke shook her head, but Bruce kept pushing.

My mind raced. Tell me what? That she had an affair with Leonardo DiCaprio during that window? That she was shooting up heroin on those days? My imagination ran wild. I reached out and held her hand. "Anything you tell me is confidential. I'm here to help."

Brooke looked at my confused face and finally nodded.

Bruce looked me in the eye and said, "We swing."

My mind flew to kiddy playground sets and wooden front porch swings and the spinning swing rides at carnivals, and I imagined Brooke and Bruce pumping away to see who could get their swing higher.

My face must not have registered the proper response, because Bruce said it again. "Dr. Rankin, we swing. Like, we *swing*. You know. *Swing*?"

Then a lightbulb flashed, just before Bruce said, "Like, we're swingers. We go to parties. We have sex with other couples."

Ah ... Okay, so I can be a little slow.

After reviewing the calendar together, we figured out, given the timing of the parties, that there was a 99 percent chance that Bruce was the father of the baby. This was followed by happy, relieved tears.

Brooke and Bruce turned out to be one of my favorite couples. What I learned from them is that getting pregnant can be a very humbling experience. You think you've made peace with your lifestyle, but suddenly you're unsure who fathered your child and your whole life winds up topsy-turvy. You question your lifestyle choices. You wallow in guilt, self-loathing, and embarrassment. But trust me. These things happen. More often than you might think. Your boyfriend breaks up with you, and suddenly you find yourself mending your broken heart with the cute guy at work who you knew always had a crush on you. Or you wind up pregnant the one time you decide to sow your wild oats and quit obsessing about monogamy.

Brooke got lucky. Because she came to see me so early, we were able to establish her conception date within a narrow window that pretty much excluded the encounters that might have landed her baby with a different daddy. But many others have no clue. If you're with Billy on Tuesday and Greg on Wednesday, your due date just isn't going to help you.

In that case, you can do one of two things. If you *must* know, DNA testing can be done on a sample of chorionic villi or amniotic fluid. If it can wait, the baby can be tested after birth. Even then, testing is imperfect.

Although some patients find themselves in court-ordered paternity testing situations we'd all prefer to avoid, many of my patients who consider paternity testing are like Brooke. They are in a relationship that may have hit a rocky spot, but they have decided to raise a baby together. As long as you are honest with your partner about any infidelity and you're committed to raising a family together, does it really matter whose DNA made your baby? At least, that's my two cents. Biological parenting and DNA can be overrated. It's love that really matters in my book.

How vigorously can I work out when I'm pregnant? I run and lift weights. Do I need to stop?

Remember, pregnancy is not a disease. It's a natural state of being for women. Which means that you're meant to keep living. If that means running for you, then by all means, keep running. I tell my patients that, with a few exceptions, they may continue doing most of the exercise they were been doing before they got pregnant. So if you're a runner, hiker, swimmer, dancer, or weight lifter, keep on keeping on. Be aware that running in pregnancy can cause stress on your joints as the baby gets bigger. And more repetitions of lighter weights are probably preferable to using heavy weights. But exercise is healthy, and I encourage you to keep doing it.

Pregnancy is not a time to pick up a new sport you've never done before. It's also not the best time to train for a marathon, add miles to your swim, or push yourself athletically. If you haven't been exercising and wish to add an exercise regimen to your life, talk to your doctor. Workouts like swimming and prenatal yoga, which includes the added mind/body/spirit dimension, tend to be a great place to start.

There are some sports that may put your pregnancy at risk, such as skiing, horseback riding, and biking. While the sport itself won't harm the pregnancy, if you suffer a serious fall, you could wind up with pregnancy complications, including losing the baby. For similar reasons, contact sports are best avoided. And you're probably best served delaying that scuba trip until the pregnancy is over.

Regardless of how you exercise, hydration is key. Drink, drink, drink water! Dehydration can trigger preterm contractions and cause you to pass out. Carry snacks to keep your blood sugar up, since light-headedness can swoop up unexpectedly and profoundly. Most of all, be kind to your body and listen to its wisdom. Pregnancy is no time to subscribe to the "No pain, no gain" philosophy. If it hurts, stop.

Does it really help if you massage the perineum for weeks before childbirth?

Perineal massage falls into that category of something you can do at home that might make you feel like you have more control over the outcome of your birth. If you lovingly attend to your yoni every day in preparation for your upcoming birth, it seems certain to bring intangible benefits. Certainly, loving energy directed at the place where your baby will come into the world strikes me as a positive action. There are mind/body/spirit connections we can't even begin to study.

But does massaging your perineum before you go into labor help reduce the risk that your perineum will tear during childbirth? No. Not according to the data.[16] However, one study did show that perineal massage performed while you are pushing reduces the risk of third-degree tears (tearing through the muscle around the rectum), although it didn't reduce the chance of first-degree (tears of the vagina only) or second-degree tears (tears in the vagina and perineum).

CHAPTER 10

Childbirth

IN AN INCREASINGLY FRAGMENTED, DISCONNECTED world, childbirth can elevate your consciousness and open a door to awakening. Although I've delivered thousands of babies, I'll never forget one special birth.

My best friend, Becca, who is also my cousin, was past her due date when the phone finally rang. In my years of carrying a pager, I became a pro at grabbing and going, but this time I wasn't going to deliver a baby. Instead, I was going as Becca's labor support, which requires more gear. Gathering up a heating pad, lavender oil, Anne Lamott's book *Operating Instructions: A Journal of My Son's First Year,* my iPod, chicken soup supplies, and some herbal tea, I raced out the door.

Next thing I knew, I was sitting with Becca in the cozy stillness of her hot tub, where the contractions seized her every few minutes. When each contraction came, I pressed my fists into her back. When the contraction passed, Becca floated, weightless and waiting, in the amniotic warmth of the pool. I felt this rush of anticipation, knowing the change that was about to take place for my best friend. I knew she would experience every rush of love, spasm of fear, and moment of transparent vulnerability that has consumed me every day since my daughter, Siena, was born. Becca's child would challenge her, adore her, and test every fiber of her humanity. She would know what true unconditional love feels like.

Later, when pain overtook her, I examined Becca's cervix. She had barely started to dilate. Becca sobbed when I told her the news. For twenty-some hours, Becca labored at home while I attended her in the presence of her doula and her husband. I did the small things that make you feel useful when someone you love is in labor. I massaged her. I fetched water. I made chicken soup. Mostly, I sat in awe of the paradoxes that surround childbirth—the strength and the fragility, the soft and the hard, the pain and the joy.

Ages later, when Becca's cervix dilated to six centimeters, we grabbed the pre-packed bags, clambered into the car, and dashed to the hospital.

Becca desperately wanted to deliver naturally, without pain medication, but her cervix quit dilating at eight centimeters. Because her doctor was slammed with deliveries, I jumped in as surrogate OB/GYN, explaining to Becca that she might need medication to stimulate her contractions. After a long, painful night, Becca opted for the epidural, and finally, with the wrenching pain numbed, Becca invited me to spoon up in her labor bed with her.

She fell asleep instantly, snoring loudly, while I curled up fetally beside her, lying on the blood—and amniotic-fluid-soaked sheets. The room was dark and still as a womb. Only the ticking of the clock marked the passage of time, and I watched the sun come up. I had never felt so in awe of Becca—or any woman, really—as I did at that moment, so skin-to-skin close, curled up on that bloody bed with my arms around her, feeling her heart beat, with her baby moving beneath my hands. Becca and I had been through so much together. We had shared too many endings and not nearly enough beginnings. But at that moment, as we lay in the bed where Becca was about to give birth, my heart overflowed with waves of love and respect.

Several hours later, the baby's head was crowning and Becca's doctor was still not there. Awash with the realization that I might be the only doctor around to deliver Becca's baby, I jumped into action and slapped on some sterile gloves. A mass of dark curls emerged from the flower of Becca's body as I squatted into my well-rehearsed catcher's position, hands outstretched. Although I had done this a gazillion times, I felt my heart leaping inside my chest. What if something went wrong? What if Becca's baby didn't take that first breath? What if my best friend began to hemorrhage and I had to start barking orders to save her life? Would I be up to the task? Could I distance myself enough to do the job, or would I collapse in a melting pile of tears?

As I held the baby's head with quivering hands, I prayed for peace and guidance. At that moment, Becca's doctor raced in and I stepped aside, breathing a deep sigh of relief. Shedding the standard blue gown I had worn so many times, I stepped back just as the baby's head spread Becca wide open and emerged like a turtle coming out of its shell. The doctor angled the baby's head downward, as one shoulder slid out, followed by the next shoulder. Then a rush of blood and amniotic fluid spattered the floor, and Becca's baby was born. Clamp, clamp, cut, and suddenly Becca was holding her daughter, laughing and crying and squealing with delight as she looked at her baby for the first time.

Wiping back my own tears, I started to sneak away, so the new family could have some private time. But Becca beckoned for me, and when I leaned in to hug her she grabbed me and pulled me into the labor bed with her, and we sobbed. That moment changed us forever. Suddenly, we were both mothers—just one more journey we would travel together.

Childbirth not only opens the door to those who share the experience but also to a whole new life for you as a woman. The changes in your body mirror the shift in your life. Suddenly, the baby emerges into the world and you are left with a body that only partially resembles the one you had nine months earlier. The vaginal tears, swollen feet, and engorged uterus heal rapidly, but other changes linger, leaving you wondering whether you will ever get your body back.

Some women bounce right back. Most of us, however, look at our postpartum bodies, emotions, sex lives, and routines and find ourselves hard to recognize. If this is how you feel, join the club, sister. Trust me, it does get better. And to be blessed with a child is worth every stretch mark, wobbly bit, and varicose vein. If we can accept that some of these changes are simply the price we pay for the joy of becoming mothers, we can save ourselves a lot of angst. So what if your body isn't quite the same after giving birth? With the right mind-set, becoming a parent helps you shift your sense of self from how you look to who you are. And trust me, baby, you're gorgeous!

In addition to being a profoundly physical experience, giving birth offers glimpses into a spiritual world where magic happens all the time. That a process as earthly as sex can result in new life baffles me still. Even though I have attended tens of thousands of births, the awe never wears off. Every time I'm afforded the luxury of being the first to hold a new baby, I am struck speechless. It's the reason we OB/GYNs endure the sleepless nights, the beeping pager, and the many sacrifices. We get to bear witness to miracles.

Is natural childbirth really worth it? I mean, I'm sure it hurts like the dickens, so why would you do it? Is it really that much better for you and your baby?

This is *your* labor experience. Keep that in mind. Nobody—I mean, nobody—but *you* can make this decision. When you get pregnant, you can be guaranteed that everyone and her mother will try to influence you. The Lamaze fans, the water birth enthusiasts, the epidural lovers, and the elective C-section crowd will all try to recruit you over to their team as if childbirth is a religion and you need to be converted. Your husband, your mother-in-law, your best friend, and your colleagues at work may swear they have the one-and-only perfect answer. Some will saddle you with detailed, gruesome accounts of their own horror stories, offering you cautionary tales of what you should avoid. Like guys telling tall tales about fishing trips, the labor pains keep getting bigger and bigger, just like the fish. Many will employ guilt tactics to make you think you're a horrible mother unless you give birth a certain way.

They even did it to me, and I'm an OB/GYN, for crying out loud! I mean *chill*, people. It's *my* life.

In spite of what others will tell you, no single labor method suits every woman. Only you will know what resonates with you, so don't let anyone try to convince you of what you *should* do. Sure, get educated and weigh your options. Seek the counsel of people you trust. But at the end of the day, tap into your intuition and figure out what feels right for you.

It's important to remember that no matter how you plan to approach labor, it is a process you cannot control. Go ahead and set goals, but release your attachment to specific outcomes. It's ultimately not up to you. Isn't that a lesson for life in general?

As for natural childbirth, I fully support those who choose to deliver naturally. Certainly, avoiding narcotics, epidurals, pitocin, and other unnatural interventions can result in healthy births and quicker recovery for mom and baby. Even more so, I think natural childbirth can be a very empowering experience for a woman, much as running a marathon can be. In both cases, you face tremendous difficulty, overcome the pain, meet a goal, and realize that if you can endure this, you can handle anything. For some women, it changes their life. It's not just childbirth—it's a spiritual experience.

But it's not for everyone—and doesn't need to be. Personally, I feel no need to run a marathon, nor do I need to experience natural childbirth to feel empowered. But that's just me. Call me a wimp, but I'm thinking, *Hmm... will I let the dentist take out my tooth au naturel, or am I gonna go for the novacaine?* Others can say what they want, but I had to make my own personal choice about how I wanted to give birth, and you have to make yours.

Is it reckless to choose an epidural? Absolutely not. Epidurals and narcotic pain relievers definitely have their risks, but untreated labor pain has risks as well. Methods we use to control labor pain are considered safe and effective, and 70 percent of women in large U.S. hospitals choose to take advantage of them.[1]

There seems to be a small but passionate faction preaching that pain during labor is normal and necessary and that drugs meant to relieve labor pain are evil. To these people I say (with the greatest amount of love and respect for your personal beliefs), if you don't want anesthesia during labor, don't get it. But otherwise, keep your thoughts to yourself when your best friend/ sister/coworker/next-door neighbor is pregnant. It's not *your* decision. It's *hers*.

Because of this underlying belief held by a few, laboring women may be treated differently than people with other types of pain. You would never suggest that someone with postoperative pain or pain from a broken bone shouldn't be given pain relief, yet you'll hear a vocal minority urging laboring women who are begging for drugs to "suck it up."

The American College of Obstetricians and Gynecologists recognizes this double standard and supports a woman's right to choose whether or not she wants pain relief during labor. Personally, I think every woman should keep an open mind, seek childbirth education, wait until the time comes, and see how it goes. You never know how you will feel until the moment is upon you. When the time comes, give yourself permission to do whatever feels right.

Am I crazy for wanting to give birth at home? I hate doctors (no offense) and hospitals give me the heebie-jeebies. Is it safe to deliver at home?

If you ask the American College of Obstetricians and Gynecologists, they will tell you're crazy. Well, not in so many words, but you won't find any support there. Their position statement recommends that birth should only take place in a hospital setting.

I, on the other hand, am not so rigid in my thinking. I think it's possible for highly skilled, accredited midwives to carefully screen women seeking safe alternatives to hospital childbirth and, if appropriate, assist them in delivering in their home or a birth center. (I can already see some in the medical community hunting me down to burn me at the stake. Quiet, doctors! Hear me out.)

Hospital birth simply does not suit everyone, because—wouldn't you know it—we're not all clones. For ages, women have given birth at home, attended by loved ones and midwives. Some long for those days, when childbirth wasn't so sanitized, medicalized, and impersonalized. I feel for those women, because the minute you walk into a hospital in labor certain rules inevitably apply and you are at the mercy of hospital policy. While the principles of autonomy still hold and you always have the right to refuse interventions in a hospital, you are likely to get the evil eye from Nurse Ratched if you try to buck hospital policy.

This makes some women insane. They don't want to fight the system in the middle of a whopping contraction. Do you blame them? These women often come to me asking for my blessing to deliver at home. As a physician, I simply can't officially endorse home birth or attend to them myself. My malpractice insurance carrier would drop me like a hot potato. But for those who choose to do so, I try to help educate them about the risks and carefully document our detailed conversation. Then I secretly whisper my blessing under the table, making sure none of the powers-that-be overhear.

So is home birth safe? Because many home births are unplanned and because many home births gone bad become hospital births, which skew the data, the evidence is limited and confusing. One study tried to make sense out of it all. This study evaluated over five thousand women who planned to give birth with a certified professional midwife at home. All patients were meticulously screened, and those deemed high-risk were transferred to the hospital. They found that the remaining home births received less than half the interventions (such as epidurals, forceps, C-sections, and the like) of hospital births. Twelve percent of studied patients had to be transferred to a hospital during labor or the postpartum period. No mothers died, but five babies died at birth, seven died in the week after birth, and two died within twenty-eight days. The overall neonatal death rate was 2.6/1000.[2] Although it's hard to compare these rates with those of planned natural deliveries in hospitals, several studies seem to support that the rates are similar to hospital births.[3]

So if medical interventions are lower and neonatal death rates seem to be the same, why does the medical community go bonkers when you bring up home birth? It's complex, so let me try to translate.

As OB/GYNs, we see the shit hit the fan, and it often comes with little warning. We'll be sitting in the doctors' lounge, numbly eating yet another bag of sour cream and onion potato chips, when suddenly an overhead page calls us to attention. We'll go racing down the hall, only to discover that a patient's baby is crashing or that the mother is suddenly unconscious. Our hearts start thumping and everyone rallies to save the baby. We race to the operating room, where we scrub our hands and prepare for an emergency C-section. Most of the time, the baby's heart rate comes back up, Mom is resuscitated, and everything's hunky-dory. But sometimes it isn't fine, and we're slicing through that belly so fast that before anyone blinks the baby is out and in the arms of the waiting neonatal specialist. All of us have seen babies and mothers die, sometimes unexpectedly. Keep in mind that these things are exceedingly rare. But once you've witnessed a father falling apart in the waiting room after you tell him he lost not only his baby but also his wife, you never, ever forget.

We like to believe that if someone is in a hospital, under our expert care, we can protect them. Of course, this is part illusion, and we revel in the illusion that we are in control. It's what we're trained to do. But the truth is, we do have some tricks up our sleeves that would be impossible to implement at home and can be lifesaving during emergencies. Because of this, I personally cannot imagine giving birth outside a hospital. Call me paranoid, but if the shit hits the fan, I want an OB/GYN at my bedside, a sterile, prepped operating room down the hall, and a pediatrician in the wings. Once again, that's just me.

So yes, while home birth seems to be as safe as hospital birth when patients are very carefully screened, transferred to hospitals readily when necessary, and attended by skilled, qualified, certified midwives, here's my question to you: If you are one of those rare women who loses a baby after a home birth, will you be asking "what if" the rest of your life? While hospital birth may be no safer, at least you will know there was nothing further you might have done to prevent such a tragic outcome.

If home birth resonates with you and you feel that you'd forgive yourself in the rare event that your baby suffers a bad outcome, I say go for it. My patients who have chosen to have home births report magical experiences that have changed their lives. Their home is transformed, as new life graces the walls of the family abode. Secretly, I wish I could be there, but alas. I must be there only in spirit, as these patients step outside the system and forge a path all their own, while I cheer quietly from the sidelines.

Does water birth really make childbirth easier? How do gynecologists deliver babies during a water birth? Are you in a swimsuit, delivering the baby? Doesn't it get really messy? Does the baby breathe in the water?

I have been blessed to attend several women giving birth in warm tubs in a hospital. The whole atmosphere was tranquil, soothing, and serene. Water has such a calming influence on me, as I think it does on many. If I had been inclined to deliver without pain medication, I would have sought out a way to deliver in warm water.

For those wishing to give birth naturally, water birth does seem to help. Epidural rates appear to be slightly lower in those who labor in warm water, and women report significantly less pain.[4] On my end, delivering the baby in water is not much different from delivering a baby whose mother is in the stirrups. The patient straddles her legs and the doctor or midwife leans over the edge of the tub with long gloves (wearing scrubs—no bathing suit necessary). Yes, the tub does get very messy, but the hospital has strict protocols for sanitizing it before anyone else uses it. And no, when the baby is born, it doesn't breathe in the water. The baby is still getting its oxygen through the umbilical cord at that point. After birth, we bring the baby to the surface right away. It's that exposure to air that seems to trigger the baby's first breath.

The key to water birth is that you must be considered low risk, since the baby is monitored only intermittently with waterproof equipment. And while some women may use warm water to soothe them during early labor before getting an epidural, delivering in the water is only recommended if you have no anesthesia. All in all, I have positive feelings about water birth. It's almost as if you're in the womb already when the birth happens.

I've heard that celebrities get C-sections just for convenience. What do you think about this?

Ooh. This would be a great instance of "Do as I say, not as I do." I'd like to say that I don't advocate elective C-section—because I don't—but that would make me a total hypocrite. Let me explain.

When I was pregnant, I had a placenta previa, meaning that my placenta was attached over my cervix, a condition that requires a C-section delivery. While some women with placenta previa are bummed out that they won't be able to deliver vaginally, I have to admit that I was relieved. Back in my twenties, I suffered from a painful type of sexual dysfunction that made intercourse feel like torture. Finally, in my thirties, this syndrome resolved and sex was fun for the first time in my life. So my secret fear about childbirth was that I would tear from my vagina all the way through my butt (a fourth-degree laceration) and sex would hurt for the rest of my life. So while I knew a C-section carried other risks, my irrational fear of tearing led me to consider elective C-section.

Ultimately, my placenta moved out of the way, paving the

way for a potential vaginal delivery. But by that time, my dad was dying of brain cancer and my life was on a downward spiral into total chaos. Frankly, I just couldn't handle one more facet of my life being completely out of my control. Motivated largely by my desperate desire for Dad to meet Siena before he died, my husband and I opted to schedule a C-section to deliver Siena two weeks before my due date, after doing an amniocentesis to make sure her lungs were fully mature.

This is not a choice I'm proud of. In fact, I lied to everybody at the time. My patients, the doctors in my practice, my best friend, even my mother. Only my doctor and my husband knew the real reason for my C-section. I lied because I didn't want anyone to mistake my choice as an endorsement of widespread elective C-section. As it turned out, Dad got to meet Siena and died two weeks later. Had I made it to my due date, he would never have met her. I have absolutely no regrets.

My exceedingly personal decision aside, how do I counsel my patients? I don't even bring up elective C-section. But patients do ask from time to time. Why would someone want to electively undergo a painful surgery? Some reasons include:

1. Wishing to avoid a long, painful labor

2. Scheduling birth in your busy day planner

3. Being delivered by the doctor of your choice

4. Avoiding going too far past your due date, which can increase the risk of complications with the baby

5. Lowering the risk of postpartum hemorrhage when compared to either planned vaginal delivery or unplanned C-section[5]

6. Fear of sexual dysfunction or urinary incontinence (although elective C-section has not actually been shown to reduce these risks)

If a patient asks, I tell her she must listen to my spiel and sign her life away in consent forms. After this, if she still wants an elective C-section, I will perform the surgery. My spiel goes something like this:

> Please don't ever choose C-section because you're afraid of the pain of labor. If this is your motivation, we can put an epidural in at the first sign of pain. And don't choose C-section because you want to schedule your birth. We can always induce labor.
>
> If you have other reasons for requesting a C-section, it's critical to understand the risks, which include:
>
> 1. Greater pain postpartum
> 2. Longer recovery
> 3. Higher risk of postpartum infection, surgical wound complications, hysterectomy, anesthetic complication, blood clots, and other postpartum complications
> 4. Greater risk that the baby will have respiratory problems at birth, especially if C-section is done before thirty-nine weeks' gestation without first doing an amniocentesis to make sure the baby's lungs are mature
> 5. Increased neonatal death rate[6]
> 6. More risk in future pregnancies, including the risk of placenta previa and accreta, uterine rupture, which may result in death of the baby, and surgical complications such as bladder injury, bowel injury, and scar tissue

As you can imagine, at this point, most women shake their heads adamantly and never bring it up again. So why would I choose to make such a risky choice? I guess you could say I'm an informed consumer. I knew exactly what I was signing up for, and my doctor respected my autonomy and allowed me to make my own decision, which is the same thing I do if a woman understands the risks and chooses to proceed anyway. It's not my job to judge anyone's choice. My job is to educate, present both sides objectively, answer questions, and respect a woman's right to make choices about her own body.

What do you think of doulas? Should I have one?

I think doulas—labor coaches trained to support the birth process—totally rock. Where I trained, I worked with midwives in the hospital, and they shaped the doctor I became as much as many of my physician teachers. I developed great admiration for the way they approach birth. After I finished my training, I moved to a hospital without midwives, and it always felt like something critical to the life force of obstetrics was missing. The way I see it, obstetrics is a spiritual practice as much as a medical practice. The doulas I have worked with seem to get this and help bridge the gap.

When doctors are running around trying to manage six laboring patients at once, you might get lost in the shuffle. Although many labor and delivery nurses are fantastic and capable of doing much of the work a midwife might do to support a laboring woman, they tend to get buried under piles of paperwork and, like the doctors, may be managing more than one patient at once. This is where doulas save the day.

A doula is a birth mentor of sorts, someone who is trained to coach you, support you, massage you, be your shoulder to cry on, and serve as your pillar of strength. While your partner may hope to be your labor coach, partners can get pretty wigged out at the site of someone they love in pain. Doulas understand the process and don't freak when you're hurting. Most doulas do not deliver babies. Instead, they nurture you through to that point, when someone else takes over for the big hoorah.

Usually, a doula attends you during every step of the labor process, in a way OB/GYNs just can't. Some even offer postpartum guidance and breast-feeding support. Is a doula necessary? Absolutely not. You will pay out of pocket for most doulas, as insurance tends not to cover them, and the cost varies widely. If you can't afford one, don't worry. Your basic needs will be met by the hospital staff. But if you do have the means, a doula is a lovely luxury, especially if you're hoping for a natural childbirth. When I find out a patient of mine has a doula I love, I breathe a sigh of relief. I know my patient will be in good hands, even if I'm too busy to be as present as I might wish.

How can celebrities have babies and six weeks later look like they were never pregnant? Is there some secret trick to Body After Baby?

Please. Do yourself a favor and don't compare yourself to Angelina/Katie/Heidi/Nicole/Halle, et al. Most of us don't look like them to begin with. And when we're postpartum—forget it. Comparing yourself to these superstars is a recipe for insecurity.

Every time I see the "Body After Baby" articles in supermarket checkout lines, I get a bit queasy. As if new mothers don't have enough pressure. Now we're supposed to look like super-models by the time we reach our six-week postpartum checkup? I mean, WTF?

Let's get real here. First, many of these women are blessed with killer beauty genes, and to top it off, they have personal trainers, chefs, nutritionists, and plastic surgeons at their beck and call. Then, if something's still not quite right in a photo, it gets airbrushed. If someone had airbrushed off the muffin top that spilled over my low-rider jeans for two years after I gave birth, I might have looked tabloid worthy, too!

Is there a trick to bouncing back quickly? I guess you could say that the magic bullet—if there is one—is not to gain too much weight in pregnancy. If you limit your weight gain to twenty-five to thirty pounds, it will be easier to wriggle into those skinny jeans afterward. But pregnancy is no time to diet, and you can't convince me that some of those celebrities are gaining as much as they should. If you look at these superstars when they're full-term, many of them still have stick legs and stick arms attached to the baby bump. Women who are underweight to begin with usually need to gain more than twenty-five to thirty pounds. If you add a pregnant belly to a stick, you're still a stick. Frankly, I wouldn't emulate many of them, if I was trying to be truly and holistically healthy. The other magic trick to losing the baby weight is breast-feeding, which increases your metabolic rate and helps you get your figure back more quickly.

But what's the rush? I tell my patients, "Nine months on, nine months off." If you still don't feel comfortable with your figure nine months out, you may need to ramp up your exercise regimen or alter your diet. Personally, I suggest that you release any expectation that you will look like a supermodel shortly after giving birth. Even if you looked like one beforehand, babies change us, and I don't think it's healthy physically or emotionally to try to regain your pre-baby shape too quickly—or at all. Learn to live comfortably in your post-baby skin, appreciate and honor what your body has done, and the beauty within you will shine forth.

Is there a time during labor when it's too late to get an epidural? What's the "epidural window" people talk about?

If you think there's an "epidural window," you've probably been watching too many movies. I've never heard an OB/GYN or an anesthesiologist use this term, but in all romantic comedies you can guarantee that the pretty pregnant protagonist misses her "epidural window" and has to huff and puff her way through natural childbirth (which, of course, makes for much better drama).

In my experience, you miss your epidural window when you're delivering in the backseat of the shiny new BMW in the hospital parking lot. (Yes, true story. And no, the shiny new BMW did not belong to the woman who defiled it. Her car was being serviced. The kind dealer had lent her his car.)

If it's your third baby and you miss your "epidural window" because you're ten centimeters dilated when you reach the hospital, chances are that you'll push that baby out faster than the anesthesiologist can race to the delivery room. But most of the time, at least at my hospital, if a woman wants an epidural and an anesthesiologist is not busy doing a C-section, she gets one, even if she's completely dilated and about to start pushing. But keep in mind that I have always practiced in metropolitan areas. If you live in Podunk, U.S. (no offense!), you might not be so lucky.

I try to counsel my patients to keep an open mind regarding epidurals. Some of my patients are adamant about getting one, while others are rabidly anti-epidural. But you never know how your individual labor will proceed. If you walk into labor and delivery at nine centimeters dilated saying, "I think I might be in labor," you probably won't be needing anesthesia. But if you're sobbing and cursing and sleep-deprived at two centimeters dilated, chances are an epidural will mellow you out and let your body do what it needs to do to dilate your cervix and allow your baby to glide into the world.

Some of my patients who miss their epidurals get pissed off, as if an epidural is every woman's birthright and our failure to deliver it constitutes malpractice. To those women, I lovingly and respectfully say, "Count yourself as blessed." Barring unpredictable and unavoidable labor and delivery mishaps, it likely means that your labor went so smoothly that you didn't have time for pain relief. Remember, we do the best we can, but labor and delivery is an unpredictable place.

Do most doctors automatically do an episiotomy? How can I avoid getting one?

I can't speak to the intention of all doctors, but personally, I don't routinely cut episiotomies (a small cut a doctor may make in the perineum during childbirth). Episiotomies are a last-minute decision for me, and if I cut one, I have a good reason. Usually, if it becomes evident that the baby will not fit through, I actually prefer letting you tear to cutting an episiotomy. Why? Because we can never quite predict what will happen in the moments around birth. Some patients we think will tear don't, and others may tear less than we might have cut. Overall, routine episiotomy is trending downward. In 1979, 61 percent of women in the United States having vaginal deliveries wound up with an episiotomy, as opposed to 25 percent in 2004.[7]

I know it sounds brutal to cut through delicate perineal skin with scissors, but we docs used to think routine episiotomy offered some protection. The rationale asserted that a surgical incision would be easier to repair, with fewer complications down the road. However, recent evidence demonstrates that we were probably misguided.[8]

While I firmly oppose routine episiotomy, it certainly has its place in some situations. This is where the art of medicine comes in. If I see signs that suggest that a woman might tear up into the urethra, I prefer to cut downward to guide an inevitable tear. But if she starts to tear downward, I'll let it go. If a baby's heart rate crashes and I have to choose between emergency C-section and episiotomy, I'll choose episiotomy, to expedite delivery. If a baby's shoulders get stuck and the baby is turning blue, you'd better believe I'm going to do everything I can to get that baby out safely, even if it means cutting. But if everything is proceeding smoothly, I'm usually better off twiddling my thumbs, putting my scissors away, and letting nature takes its course.

This is why it's important to choose a doctor you trust. These are split-second decisions we make. You have to know that your doctor has your best interest at heart.

Do I need to write a birth plan, or should I trust that my doctor knows what to do?

If you say the words "birth plan" to a group of OB/GYNs, you may see us cringe and roll our eyes. Why? Here's a typical scenario from a doctor's perspective:

Ava is thirty-two weeks pregnant with twins, and she is booked for a seven-and-one-half-minute appointment in your already-overcrowded schedule. You've spent fifteen minutes with her already, which means that your next three patients will be pissed off because you're running behind. Just as you finish listening to her baby's heartbeat, Ava hands you her birth plan and asks you to take a look. It is twenty-three pages long. The first ten items read like this:

1. I will be bringing two musicians whose drums will guide my babies' spirits into the world. I will need a guarantee that they will be permitted into the delivery room.

2. I do not want an IV under any circumstances.

3. I do not want to have my babies monitored. I am intuitively in touch with them, and if something is wrong, they will alert me.

4. Please do not restrict my intake of food or drinks. I intend to feed my body nourishing foods as it experiences this natural process.

5. I will scream bloody murder if the nurse even mentions the word "epidural."

6. My seven-year-old daughter wants to deliver the twins, so while we trust you, we expect you will let her do the actual deliveries.

7. I do not, under any circumstances, want a C-section.

8. When the first baby is born, we want the umbilical cord to remain attached until the placenta is delivered.

9. When the second baby is born, we want both babies simultaneously brought to my breasts so one twin doesn't feel more bonded than the other.

10. When the placentas are delivered, we want to save them. We will be bringing a blender into the delivery room so the placenta can be turned into a milk shake I will drink for nourishment.

(And this is just the first page.)

You think I'm kidding. You're probably giggling and gagging right about now. I swear to God, I'm not kidding. Suddenly, you realize that you will be running at least an hour behind, not to mention that Ava is *not* going to be happy with what you need to say.

Such efforts to control an uncontrollable process have given birth plans a bad rap in the eyes of many a doctor or midwife. That said, I think birth plans can be a fabulous launching pad from which to have a heart-to-heart conversation about your personal childbirth wish list. But keep that in mind—it's just a wish list. While I try to honor the requests on each woman's list, my number one priority is always "healthy mom, healthy baby."

If you want your doctor or midwife to be receptive to your birth plan, consider starting it with a statement that sets the tone for cocreation and collaboration during the birth process. Something like "I know we can't predict the future, and we trust that you will do whatever is necessary to help us have a healthy childbirth, but if possible, we would love it if [x, y, z] could happen." Rather than putting your health-care provider on the defensive, this will invite a spirit of trust and recognition that *birth* is simply not something we can *plan*.

Is a birth plan necessary? Absolutely not. My friend Amelie hired a doula to help her through her fourth delivery so that her poor husband, who simply does not like blood, could be off the hook. Her doula asked her to write a birth plan, which Amelie thought was ludicrous. After all, she had given birth three times. When her doula insisted, Amelie wrote: "Go to hospital. Get epidural. Have baby. Drink port." Her doula was not amused, but I thought it was friggin' hilarious.

So no. A birth plan is not necessary. Sure, if you would like to communicate your desires to your health-care providers, a birth plan can help. But do us a favor: Be reasonable. Assert your wishes, but release any fierce attachment to having those wishes come true, and bring up the birth plan at the beginning—not the end—of your prenatal visit. And if things don't go as planned, let it go. As long as you and the baby are healthy, you have been blessed.

How can I know if my vagina is big enough to deliver vaginally?

You can't. If you try to deliver a baby vaginally and your baby doesn't fit through, that means that this particular baby, in this position, during this labor, doesn't fit. But the next baby in your next labor might fit perfectly. On the flip side, your first baby might slide out like a greased watermelon in a swimming pool. But that doesn't mean your second baby will. Doctors used to employ techniques like clinical and X-ray pelvimetry—ways to measure the pelvis to assess whether it was adequate to allow a baby to fit through. However, these tests are not consistently able to predict whether a woman will be able to deliver vaginally and are hardly used anymore.[9]

The best way to find out whether your baby will fit is to try to deliver vaginally. The only time we bring up the subject is when a baby is exceedingly large on ultrasound (more than forty-five hundred grams—about nine pounds, fifteen ounces—is usually the number that starts the conversation). In these cases, I usually discuss the possibility of C-section, because the risks of shoulder dystocia (when the baby's shoulder gets stuck), inability to deliver vaginally, and birth trauma rise. Even then, many women deliver little sumo wrestlers without a hitch. You never know until you try.

How come I still look totally pregnant two days after giving birth? It's not like I thought I'd be bikini ready, but come on already. What's the deal?

Oh, girlfriend, I could show you some pictures that would make you feel better. When I looked at my naked belly in the shower two days after my C-section, I was convinced I must have conceived twins. I asked my nurse, "Are you sure they didn't leave another baby in there?" No kidding. I had packed a cute Mommy-goes-home outfit but had to send my husband home to bring my maternity clothes, which I used to cover up as I tried to slink out of the hospital before my colleagues saw me and asked, "So,

Lissa, when's the baby due?" I knew my fragile ego just couldn't take the blow.

A week later, I was at the farmer's market, gorging on unpasteurized cheese samples after nine months of abstaining. The cheesemaker took one look at me and said, "This cheese is not pasteurized." I nodded and kept right on eating. He kept looking at my gut, then making eye contact, then looking back at my gut.

"It might not be good for you," he said. I knew where he was going with this (unpasteurized cheese can harbor harmful bacteria that can hurt a pregnancy), but I couldn't resist making him squirm.

"Then why would you be feeding it to all these people?"

"It's fine for others. But it might not be fine for you."

I couldn't bear it anymore. I broke down and told the guy I wasn't pregnant. He turned beet red and gave me a hunk of free cheese.

Why does this happen? Why did I—and why do many of us—still look thirty weeks pregnant after our babies are born?

After giving birth, your uterus, which is normally tucked neatly down in your pelvis, still hangs out way up by your belly button. So a bulge can be expected. Plus, chances are that your bowels haven't moved in days and your guts are bloated up like balloons. Then there's all that water weight your body is processing, since it takes the 50 percent more blood you had in pregnancy and shunts it all over the body before passing it through your kidneys. And then (yes, it's true), there's probably a wee bit of baby fat still left right there in your midsection. But don't despair. Give it six weeks and you'll be well on your way to getting your body back.

Does having kids really stretch out your vagina?

Yes, if you deliver vaginally, it usually does. And if you have a ten-pound baby or four kids, it may be even more stretched out than normal. How elastic your tissue is depends on many factors: genetics, whether you smoke or have other health problems like diabetes, how frequently you do Kegel exercises, age, and hormone status, to name a few.

I can usually tell whether a woman has had children when I perform a speculum exam. Women who have had children may require a bigger speculum, while those who have not often feel tighter when I examine them. For some women, the stretching caused by childbirth is a blessing. If sex has been a tight fit with your partner, leading to painful intercourse, having a baby may make more room and allow sex to be more pleasurable. And if Pap smears used to hurt, they might not anymore. But some couples complain that the vagina stretches out so much that sex feels less stimulating. If this is the case, Kegel exercises can really help. If this fails to help and if you have other symptoms of pelvic prolapse, talk to your doctor. She might be able to help you by performing surgery or inserting a pessary, which is kind of like a diaphragm you stick in to help hold things up.

I just had a baby and my vagina's falling out. What's the deal?

I hate to break it to you, but you probably have pelvic prolapse, just like my friend Aria. I was at a dinner party with Aria, whose enormous baby I had delivered six weeks earlier, when she piped up, "Lissa, I've got a droopy box. Come look!" She dragged me back into the bedroom, dropped trou, and sure enough, the girl had a droopy box.

If you just had a baby and you feel like your vagina's falling out, you're not alone. Postpartum mothers, especially those who are breast-feeding, may suffer from pelvic prolapse, which may be temporary but can be permanent. Because estrogen levels are low when you're breast-feeding and pelvic tissues may not be done healing, many women notice laxity in their vaginal area, which may result in a noticeable bulge at the introitus (the opening of the vagina). You may also leak urine, which just adds insult to injury. As if pregnancy and childbirth aren't big enough blows to the maternal ego, nature has to add a saggy cooch to the mix. But look what you've got to show for it. Trust me, it's worth it, but if it's bothering you, see your doctor.

Should I make sure my doctor waits until the cord stops beating before the placenta delivers, or is that an old wives' tale?

No, it's not just an old wives' tale. While many doctors clamp and cut the umbilical cord immediately after birth out of convenience, it appears to be true that delaying cord clamping for a few moments after birth improves outcomes. After the baby is born, blood from the placenta transfuses into the baby via the umbilical cord, especially if the baby is held below the level of the placenta. According to several studies, when cord clamping is delayed, babies are less likely to be anemic and need transfusions. Preterm babies with delayed cord clamping have the additional benefit of lower rates of bleeding in the brain.[10] The way I figure it, why not give babies every shot they can get?

I've heard that some women eat their placentas after childbirth. Ew! Is this true, and why would anyone do that?

Yes. Some women do eat their own placentas, either raw or cooked, after childbirth. While this practice, called "placento-phagia," is uncommon in humans, I have actually had patients ask to be allowed to eat their placentas. As far as I'm concerned, it's their placenta and they can do what they please with it.

Why would you eat a placenta? Rumor has it that it helps prevent postpartum depression and may help contract the uterus after birth, but I'm not aware of any studies to support this. Animals are known to eat their placentas, which may help supply necessary nutrients, but a normally nourished human is unlikely to need this type of nourishment.

If you're tempted to eat your placenta, but the idea makes you queasy, consider other placenta rituals. Some families save the placenta, bury it in the ground, and plant a tree or flower over it. Supposedly, the placenta enriches the soil and draws the life force from the earth into the plant, which celebrates the birth of the child. I'm a big fan of rituals and how they can honor important transitions in your life.

How come I lost my sex drive after having a baby?

For nine months, your body rages with hormones. Estrogen and progesterone course through your bloodstream, plumping up your sexual tissue and, for many, revving up your libido. But sex gets tricky in your third trimester, so you may be counting the days until you can enjoy sex without a big belly between you and your partner.

Then *bam*. The baby comes, and you find yourself sitting on a blow-up donut pillow for two weeks until your vaginal tear heals. At the same time, your breasts explode into milk factories and your hormones plummet. Add to that chronic sleep deprivation and shifting dynamics between you and your partner. Oh, and there's likely to be a baby sleeping where you usually make love. It's no wonder most postpartum women find their love life lacking.

One friend whose baby I delivered said, "I honestly don't care if I never have sex again." I've heard that comment time and time again. For most people, this feeling is fleeting. When you close up the milk factory and your baby starts sleeping through the night, you may feel differently. As your body heals, your hormones return to their pre-pregnancy levels, and the dust from the new-baby fallout settles, you're likely to discover a burning ember in the dying fire of your libido. If you stoke the fire, breathe life into it, and prioritize your love life, you'll be able to rekindle the flame.

Now that I've given birth, sex hurts like hell. I'm six months postpartum. Will it ever get better?

Two of my close friends couldn't have sex without excruciating pain for over a year after giving birth. Both were first-time moms, and both tore during childbirth. One had a more serious third-degree tear, through the muscle around the anus. The other had the mildest first-degree tear, injuring only the mucosa of the vagina. Although their tears differed, their pain afterward was similar.

In both cases, two factors were at play: The wound in the vagina and the vaginal dryness related to breast-feeding. Both had delayed healing of their tears. Rarely, the process of healing gets arrested, delaying wound healing. By six weeks postpartum, most tears will be completely healed. If not, we docs have some tricks up our sleeves to help finish the process, but it may take some time.

Once the wound heals, the other factor plays in. Because estrogen levels fall so sharply when you give birth and start nursing, the lining of the vagina can become very thin, dry, and fragile, making sex hurt. Using sexual lubricants while you're nursing can help. If your dryness is severe, talk to your doctor about whether estrogen applied locally in the vagina could help.

And yes, it does get better. Wounds heal. Estrogen levels rise when you stop nursing, and, barring some rare complication, your yoni will hit its stride once again. If time has passed and sex still hurts, see your gynecologist. You may have developed a gynecologic condition that needs treatment.

The good news is that things tend to go more smoothly the more babies you have. You're less likely to tear if you've had a baby before, and because the vagina tends to stretch out with multiple births, the vaginal dryness caused by nursing may not bother you as much the second time around.

I just gave birth and I'm overwhelmed with feelings of regret. My whole life has changed and I miss my old life. I can't tell anyone how I feel—they will think I'm a horrible mother. What's wrong with me? Where are all those maternal instincts they talk about?

I completely understand how you feel. Before the baby, you likely enjoyed eight hours of uninterrupted sleep each night, a body that felt like yours, the ability to focus on your work, the undivided attention of your partner, and personal time to rejuvenate, relax, and pursue your passions. Now all of a sudden, those luxuries feel like a thing of the past. And to top it all off, you're responsible for this helpless little person who is a bawling ball of need, unable to even thank you for all your sacrifices. To make matters worse, you may feel like you're supposed to be instantly overwhelmed with feelings of love, tenderness, and joy for this creature you just brought into this world. After all, that's how it is in the movies, right? You give birth, then your whole life is rose petals and fairy wings, right?

While they may not admit it, not even to themselves, I would put money on the belief that every mother has felt exactly like you at some moment in her maternal life. Having a child changes you. When you give birth, you also experience a sort of death—the death of life as you knew it. It's only natural to mourn that loss of the self you know and love. It's important to give yourself permission to grieve.

This doesn't mean you can't reclaim your self, rebuild your life, and emerge from your ashes as someone even better. But this doesn't happen overnight. There's a process we all go through, if we're honest with ourselves. The first few months pass in a blur. On one level, every moment of being a new mother feels eternal—the tedium of night feedings and pediatrician visits and shaking the rattle over your newborn's blurry eyes. On another, it passes in a blink. But over time, the grieving for your old life passes and you step into gratitude for the new life you've created.

When you choose to become a parent (or if circumstances choose you, as the case may be), it's like standing at a crossroads. One road leads to a childless life, with all its joys and sorrows. If you choose that path, you may travel more, experience more adventure, retain your girlish figure, and enjoy more freedom. But later in life, you may long for the quiet joys of family. On the flip side, if you choose to take the path of motherhood, you sacrifice much. But you'll likely be prone to moments of elation that you might have never known if you hadn't become a mother.

Patients have asked me whether I "recommend" becoming a mother. I tell them that I don't buy into any of the romance of having a child. Personally, I could have gone either way. Had I chosen to be childless or had the Universe denied me the opportunity, I would have lived a very different, but certainly blessed and happy, life. Because I chose to take the other fork in the road, I can't imagine my life without my beloved daughter. But it's not black or white, better or worse. It's just different.

There's no point looking back. You have embarked upon the motherhood journey, and I guarantee that you have within you everything it takes to rise to the occasion. It's all about how you respond to the change. To quote my friend Dr. Joanne Perron, who is an OB/GYN, yoga instructor, and sage, "Pain is inevitable. Suffering is optional." Joy is a choice. Feel your feelings and know that it's okay. But don't fall into wallowing.

Keep in mind that feelings of inadequacy, doubt, and regret are common and normal. But the line that divides normal from postpartum depression can be fuzzy, and it's critical to figure out where you lie in relation to this dividing line. How come many women feel just like you, but only a desperate few go on to neglect or hurt their children or themselves? Depression can devastate.

To figure out where you stand, check in with yourself. Can you still experience moments of joy? Do you feel hopeless, helpless, and worthless? Can you look ahead and see a light at the end of the tunnel or do you feel like you're in a dark, dead-end cave of despair? Trust the truth that lies within you.

If you're unclear about whether your feelings are normal, talk to a therapist or physician, who can screen you for postpartum depression. If you're considering hurting yourself or your children, get help immediately. These feelings are nothing to mess around with, and the sooner you can process the change in your life, the sooner you and your baby will thrive.

Why do some women get postpartum depression and others don't?

Nobody knows exactly what causes baby blues or its more severe form, postpartum depression—and 40 to 80 percent of women develop postpartum blues. They include temporary mood changes (wild swings from elation to mournful sadness), testy irritability, overwhelming anxiety, feelings of doubt and regret, and fits of weeping. These feelings typically peak in the first two weeks after giving birth and tend to improve after that.[11]

I can attest to the timing of this. Typically, if women deliver vaginally, we see them back for a postpartum visit six weeks after the birth. But if women deliver via C-section, we see them two weeks later to check their incision. Those who come in at the two-week mark are noticeably fragile. If I hold a woman's hand, look into her eyes, and ask how she is doing, the question usually triggers a wave of tears. When I see the same woman back at the six-week mark, she's usually figured out how to cope. Not to suggest that it's in any way abnormal, but those women who are still sobbing at the six-week mark catch my attention as mothers needing further assessment.

Why do some women get postpartum depression, while some skip the blues altogether? We haven't figured that out yet, but there are some well-known risk factors. You're at higher risk if you have a history of depression, your marriage is in disarray, your pregnancy was unplanned or unwanted, you've experienced significant life stressors, your partner and family aren't supporting you emotionally or financially, you're a single mother, you're unemployed, your baby has a disability, or you've lost your baby.

Even still, some women have none of these risk factors and suffer desperately. Fortunately, there's treatment, so tell your doctor if your mood changes linger or are profound.

I opted not to find out the sex of my baby beforehand, but I hadn't realized how desperately I wanted a girl until the doctor said, "It's a boy!" and my heart sank. What's wrong with me? Am I the world's worst mother?

No, honey, you're not. In fact, your feelings are so common they even have a name—*gender disappointment*. When you saw that little newborn penis, you began to mourn a whole host of dreams—buying your daughter puffy pink dresses, braiding her hair, helping her plan her wedding, chatting over herbal tea at fancy spas, and attending the birth when she delivers your granddaughter. Many women long to raise a child of a particular gender and may not even realize how profound their longing is until the baby with the wrong genitals lies in their arms.

For women in the United States, gender disappointment occurs most often when a boy is born. As women, we may see raising a daughter as more natural, as if that XX chromosome links us in some primal way that will make becoming a parent easier. While playing dress up, throwing tea parties, and nurturing baby dolls may seem second nature to us, the thought of rowdy little bruisers growling at construction equipment and racing choo-choo trains right through a little girl's Barbie Dream House may scare the hell out of you. Raising a boy conjures images of violent video games, headbanger music, and football practice, something that may feel much more foreign to you. It's no wonder you feel sad.

In other cultures, the opposite may be true. If you are expected to produce an heir, or if you live in a society that values males but not females, you may mourn this loss, too. But mourning a dream doesn't mean you won't love this child and be a glorious mother. You may hear phrases like "as long as the baby is healthy," which exacerbate your feelings of inadequacy. But don't let others invalidate your emotions. Allow yourself to feel your feelings, and don't get caught up in guilt, shame, or regret. It's only natural to feel sad when a dream doesn't come true. Harboring your disappointment in secret will only put you at greater risk of conditions like postpartum depression. Feelings are not right or wrong—they just *are*. If you need to, seek out the help of a therapist skilled in dealing with postpartum issues. But don't fret. You will move through your feelings and find all the love you need within you to raise a happy, healthy child, regardless of the gender.

CHAPTER 11

Menopause

I FOUND MYSELF PROCRASTINATING ON this chapter, saving it until the very last, and I couldn't figure out why I was so hesitant to write about menopause. After all, more than half of my patients are menopausal and I've spent more than a decade teaching women about this natural evolution in the life cycle. Finally, I figured it out. Of all the main topics of the chapters in this book, menopause is the one thing I haven't personally experienced. I am always reluctant to speak as if something is true when I haven't experienced it firsthand. I know men write with expertise and wisdom about childbirth and menopause, periods and vaginas without experiencing any of these things. But I've always dug deep within when I write, so it's harder for me to write about what I haven't yet experienced.

I know my day will come. I used to dread menopause, as many of my patients do. Who wants to endure hot flashes, night sweats, mood swings, brain fog, insomnia, vaginal dryness, skin changes, and weight gain? But as this life transition draws closer, I find myself shedding my fear and beginning to embrace it. I would never want to move backward in my life. Each year is filled with lessons learned and growth experiences. God forbid I should ever have to relive my twenties, full of insecurities, vanity, pride, and selfish choices. And my thirties, while vastly better than my twenties, were largely a time of walking through life like a zombie, asleep to my true calling. Just this year I have embarked upon my forties, and already I feel different. I'm getting smile lines and gray hair and age spots, but I really don't give a flip. Sure, the radiance of youth may fade as we age, but we glow with a different kind of light as we step into who we are truly meant to be.

Because of this, I no longer dread menopause. Instead, I see the hot flashes, mood swings, and insomnia as part of the transition. Just as the pains of labor and childbirth allow us to create a whole new life, the discomforts of menopause usher in a rebirth of another sort. Many of the women I most admire have passed through this phase and emerged more fully as confident, gifted, loving, secure, beautiful human beings. Someday, that will be me.

In bygone days, we weren't expected to live long past menopause. Even when our life expectancy began to increase, menopause was viewed as the beginning of the end. But not anymore. These days, the average age of menopause is fifty-one and women in developed countries are expected to live, on average, eighty years or more. Menopause is no longer considered a time of endings. In fact, the journey is just beginning.

It seems fitting that the word *menopause* has within it the word *pause*. After all, it does represent a shift. Menopause can be a rich time of growth, an opportunity to pause, to regroup, to change. It represents the transition from fertile young woman into mature, wise sage.

Men are afforded no such obvious change. Perhaps if they were, they would be less inclined to race around with pretty young betties in overpriced sports cars while grasping their toupees. The way I see it, menopause is nature's cue to slow down, to take stock, to assess the life we've lived so far and set intentions for the life we have left to live. I've witnessed thousands of women in my practice transition through menopause, and the one thing I've learned is that menopause is what you make of it. Menopause is a state of mind.

Even so, menopause feels like a loss to women like my patient Marianne. Her whole life, Marianne has been valued for her beauty. As a teen she was head cheerleader, and she began modeling in her twenties. For most of her life, she acted in TV commercials and the occasional bit part in movies. In the beginning, she played the sexy young girl, then, as she aged, the hot mama. When she turned forty, she underwent what she called her "forty-thousand-mile makeover," enduring multiple plastic surgeries to maintain her beauty.

Menopause hit Marianne hard. With her entire sense of self wrapped up in her youth and beauty, Marianne fell into a downward spiral of depression that coincided with her falling hormones. Hot flashes, insomnia, and mood swings plagued her, and hormone replacement therapy failed to help. Nothing I could say or do kept her from clinging to her youth like a drowning woman grabbing a life preserver.

My patient Lily approached menopause differently. As she approached menopause, Lily found herself yearning for the cyclic life rhythm her menstrual cycles brought to her. Her menopause coincided with the end of a romantic relationship and her only son going off to college, so it was already a time of mourning. Sensing the need to continue cycling, even in the absence of menses, Lily looked to nature for the cyclic rhythms she craved.

Lily is an avid gardener who tends orchids and loves stargazing. One night, she conceived an idea for a new monthly ritual. Now that her menstrual cycles no longer remind her of the monthly possibility of life, she marks the rhythms of her life by the phases of the moon. During different lunar phases, she takes her orchids outside for a moondance, holding them up in the moonglow, twirling about. Her orchids represent a monthly rebirth, a transition from bloomless stalk to delicate flower. While she can't bear children anymore, she can still bear orchids. Celebrating them in the light of the moon fills her with a sense of monthly connectedness, with both her body and Mother Earth. From this ritual, she fulfills her need for the cyclic energy she craves, ameliorating the loss she experienced with menopause.

Lily's story filled me with awe, and it wasn't just the breathtaking visual of this lovely woman moondancing with orchids. It was because she was so in touch with what she needed to feel whole. I suspect that her moondance has something to do with the fact that Lily's menopausal symptoms have been much less severe than Marianne's.

My friend Catrina saw menopause as liberating. No longer concerned about birth control or periods, she was finally able to focus on the career she had long ago put on a back burner so she could raise her family. As her hormones changed, Catrina noticed something else shifting, too. She no longer felt the need to put everyone else's needs first. When her teenage kids demanded things from her or her husband complained about something around the house, she found herself standing up for herself and placing new value on her time and energy. She said, "It's as if shutting off estrogen flipped a switch that turned on *me*."

As your body shifts away from the reproductive hormones associated with fertility, your priorities may change, too. You may find yourself less focused on being the caretaker and more interested in exploring the world outside your home. Dreams long dormant, passions quelled, education postponed, and careers neglected may begin to whisper in your ear. Your role in the family may shift, which may spill over into other areas of your life. No longer the damsel, you transform into the mythological archetype of the crone. Some of my friends even celebrate with crone ceremonies, honoring the woman they have been and rejoicing in the woman they are becoming. While levels of estrogen and progesterone fall, you may find yourself mourning the loss of your reproductive years. But have faith. You may be gestating something new, a seed within you that's getting ready to sprout. Nurture that seed, tend it, and allow it to blossom into who you really are. It's time to celebrate!

How do I know if I'm in menopause?

By definition, you have officially hit menopause one year after your last period or when your ovaries are surgically removed. While this is the technical definition of menopause, it's not a very practical one, since estrogen levels begin to fall long before you've skipped a year's worth of periods. Many women experience symptoms of estrogen deficiency even sooner than that. The term *perimenopause* comes in handy for defining that in-between time, when your hormones are declining but you're not technically in menopause yet. If you've had a hysterectomy, you're taking birth control pills, or you have a condition like fibroids that makes you bleed, it may be harder to tell when you've hit menopause.

Most women know when they are approaching menopause because their hormones go cuckoo. Estrogen withdrawal can lead to embarrassing hot flushes, pajama-soaking night sweats, sex-busting vaginal dryness, relationship-threatening mood swings, irritating forgetfulness, and changes in skin firmness that leave you thinking about Botox. It's enough to turn a perfectly sane, capable woman into a basket case. Granted, some breeze through menopause. However, if you are one of those basket cases, see your doctor. We can help.

I thought menopause happened to women in their fifties, but I'm in my early forties and my doctor told me my crazy heavy periods and mojo-busting mood swings are because I'm "getting close to menopause." What the hell is up with my body?

Our hormones begin to shift long before we are technically in menopause. Women in their forties may feel just like newly menstruating teenagers who have wild hormonal shifts that result in heavy, irregular bleeding and moody outbursts. Bloodstained white skirts may leave you flashing back to junior high school, only this time you're in the boardroom when the scarlet stain spreads across your backside.

Why does this happen? At the extremes of your reproductive life—in your teens and in your forties—you tend not to ovulate, which means that nobody's captaining the ship of your body. Without hormones charting your course, you bump and sway all over the map. You might have high estrogen levels one day that bottom out a day later. The balance between estrogen in your body (which makes the uterine lining grow) and progesterone (which tempers its growth) may shift, leaving you with a relative estrogen surplus, even as estrogen levels begin to fall off. These hormone fluctuations may turn your periods into blood floods and leave your mood careening up and down like a seesaw.

We call this often prolonged hormonal roller-coaster ride "perimenopause." While there are ways to diminish symptoms that accompany perimenopause and your doctor can help, try to go along for the ride rather than fight the inevitable. The ups and downs are just part of the natural changes our bodies undergo, as we get ready to evolve into the next phase of life. Go with the flow, and make sure you throw those arms up in the air while you're thrill riding.

I was thinking menopause would be over and done with in a year, but it's been five years since my last period and I still feel like a raging menopausal beast. When does menopause start and how long does it last?

When it comes to menopause, women are as different as fingerprints. You simply can't generalize. One woman might simply stop menstruating without a single symptom. Her sister might be plagued by hot flashes, night sweats, mood swings, insomnia, and sexual issues for a decade or more. Your symptoms may depend on what your baseline estrogen levels were like for most of your reproductive life.

Estrogen levels vary widely from woman to woman. For example, some women normally walk around with very high estrogen levels. These women tend to have curvy, voluptuous, Rubeneque body types. Other women may have normally lower estrogen levels. Think tall, slender, small-breasted marathon-runner types. When the Rubenesque woman goes through menopause, she's used to having much higher levels of estrogen, so menopause might hit her hard. On the flip side, the marathon runner, who has been hanging out with lower estrogen levels, may barely notice a drop-off in her hormones and glide right through the change.

But many other factors play a role—genetics, diet, lifestyle, exercise, mind-set, stress levels, and cultural perspectives. There's no way to predict what your menopause experience will be like. I do believe you have some control over it, though. Approach menopause as an opportunity, not something to be dreaded. Keep a positive, open mind and look for the riches in the experience. While you're mining for the pot of gold, you just might find a rainbow.

I've been through menopause, and now my vagina itches so much I could take sandpaper to it. Why?

While vaginal itching before menopause most commonly represents a vaginal infection, things change after menopause. Not that yeast infections don't still happen, but during menopause other causes of itching emerge to torture the poor, innocent coochie. Between puberty and menopause, the vagina lives in a steady state of pink, fluffy, moist, resilient health. As estrogen wanes in menopause, the vagina may become thin, dry, red, and fragile. After menopause, the inside of your vagina may wind up itching, burning, and dry due to *atrophic vaginitis*. Before you take sandpaper to your va jay jay, see your gynecologist. Estrogen aimed at the vagina can improve your vaginal health without putting you at risk the way systemic hormones may.

I'm in perimenopause, and I honestly think I'm going wacko. I used to be this even-keeled woman, and now, I swear to God, my husband is ready to divorce me. I'm always yelling at him and saying the kinds of things I never would have said in the past. Is this what they mean by a midlife crisis?

I doubt you're going wacko. Chances are you're experiencing some mood swings as the result of fluctuations in your hormones. And sometimes, when that can of worms gets opened, everyone had better get out of the way. Menopause tends to be a time of empowerment for women. Perhaps you've been the quiet, obedient wife for all of these years and suddenly you're speaking your mind. Your husband may not be prepared for the changes you are undergoing.

Some of these changes may be temporary. When your hormones settle down, or if you choose to take hormone therapy you may notice that your emotions become more even-keel again. Or they might not. Sometimes these hormone fluctuations liberate us, giving voice to long-unspoken thoughts and feelings. Divorce around the time of menopause happens quite a bit, but that's not always such a bad thing. Sometimes it takes a profound change to shine the light on a relationship that hasn't been working for eons. Suddenly, with children growing, priorities changing, and roles evolving, marriages may not survive the upheaval.

As for whether you're having a midlife crisis, I can't say. But yes, I suppose many women in midlife experience a crisis of sorts. Any life change may precipitate it. For me, it was losing my father within two weeks of giving birth that pushed me into my midlife crisis. For some, it's menopause. That said, I don't believe a crisis at midlife is necessarily a bad thing. More often than not, it's a time to regroup, to reset priorities, and to take stock of our lives.

I find it alarming how many women get put on antidepressants during menopause, as if we need to be medicated out of how we're feeling. Most of us don't need drugs. We need permission to have a voice.

If perimenopause is causing marital discord for you, consider seeing a marriage therapist. If you feel your mood swings are hormonal, talk to your doctor about ways you might stabilize your mood. And most of all, give yourself permission to be present with what comes up.

Now that I'm in menopause, is it inevitable that I'm going to be a fat cow?

Hey, go easy on yourself, girlfriend. Don't be calling yourself "a fat cow." I prefer the term *fluffy*. But I hear what you're saying.

While it's not inevitable, many women do experience weight gain in menopause, unless they change how much they eat and exercise. Because your metabolism slows down, you may gain weight if you don't eat less or exercise more.

You might be one of the lucky few who can keep piling on the carbs without ever moving your booty, but those women are rare. The way I see it, menopause is a good time to reassess your health habits. As we age, eating well and exercising regularly are just part of building better lives. Let this be a time to optimize your wellness in anticipation of a long, full, vital life.

Is it true that the uterus can fall out completely?

Yes. I've seen it with my own eyes.

One patient, Gloria, called me late at night, screaming bloody murder. "I've got a piece of meat coming out of my privates!"

Since I've seen any number of things coming out of vaginas over the years, I had to ask, "Did you put any meat inside of you?"

You'd have thought I asked her if she had sex with a skunk.

"What kind of crazy lady do you think I am? No, I didn't put meat inside of me. But you better believe that meat is coming out."

When I asked her to describe what she saw, she said, "It looks like a piece of pork tenderloin, only part of it is shaped like a donut hole." Right away, an image flashed into my mind as Gloria said, "Doctor, I have to tell you. I am really *freaking out* right about now."

I agreed to meet Gloria at the emergency room.

Sure enough, just as I suspected, her uterus had fallen out of her vagina and was hanging between her inner thighs. Turns out that this seventy-eight-year-old woman had been helping her son move a sofa when suddenly she felt a plunk, and *presto!* Meat between her legs.

Usually, this doesn't happen quite so suddenly. The uterus begins to prolapse gradually. First it falls into the vagina, then the cervix might peek out from the vaginal opening, and over time it may fall all the way out (a condition we call "complete procidentia").

In Gloria's case, I was able to push her uterus back inside and secure it with a pessary (a device that looks like a diaphragm, only sturdier). A few months later, a hysterectomy and a little vaginal nip and tuck solved the problem. But Gloria is still recovering from the trauma of what she calls "the day I gave birth to a pork chop."

Do little old ladies still have sex?

Absolutely. While some older women choose to close up shop down there, many are still open for business—and thriving. Let me tell you a story.

When I was a brand-new OB/GYN, I was seeing patients in the continuity clinic, where we young docs got to practice having our own clinic. Mrs. Kaufman was my very first patient, and I'll never forget her. She was eighty-four years old, and the words scratched on the top of her chart read: "Patient complains of vaginal itching."

Before I entered her room, I stood outside her door, scrolling through the differential diagnosis for vaginal itching in my tired, stressed brain. Yeast infections, bacterial vaginosis, trichomoniasis, atrophic vaginitis, psoriasis, vulvar cancer, warts. Could an eighty-four-year-old actually have warts?

Mrs. Kaufman looked like a storybook grandmother, clad in a hand-knit lavender sweater, long white hair knotted neatly into a bun, with bifocals on a string around her neck. "Hi, Mrs. Kaufman. I'm Dr. Lissa Rankin," I said, stuttering over the title and fumbling through her chart.

"Well, hello, Dr. Lissa," she said, gazing up from her knitting and smiling.

Doing what I'd rehearsed in my head, I shook her hand and sat on the rolling stool next to her. "What brings you to see the doctor today?"

She stood up and put her hand on my shoulder, looking me right in the eye. "Honey, I got me an itch."

I nodded, making notes in the chart, wondering if she could tell by looking at me that I felt like I was play-acting the doctor thing. Though I had jumped through all the hoops like a poodle at a pathetic, road-weary traveling circus, I definitely didn't feel like a real doctor yet. I went through the motions, sitting up straight and sticking a pen in my tied-up hair, while I asked the required litany of questions: "How long have you had the itch?" "Where exactly is the itch?" "Does anything make the itch better or worse?" "Have you tried any home treatments?" "Have you ever had an itch like this before?" I furrowed my brow, trying to look serious and doctorly. Mrs. Kaufman answered all my questions.

Handing her a gown, I asked her to get undressed from the waist down while I stepped outside the exam room for a moment. I snuck out the door, feeling my heart race. I had never done a gynecologic exam without someone looking over my shoulder.

When I walked back into the room, my expression must

have betrayed my feelings, because Mrs. Kaufman patted me on the shoulder, as if she were the doctor and I were the patient. Awkwardly I leaned over and pulled out one rickety metal stirrup that clicked and creaked as I tried to adjust it. Seeing me struggle, Mrs. Kaufman stood up from her chair, steadying herself with her cane, and pulled out the other stirrup for me, getting it just right the first time.

I was about to give her my arm to help her onto the table when she put both hands on my shoulders and said, "Sweetie, I gotta tell you something first." I was sure she was going to tell me she was on to me—that she could see right through my skin to the insecure child I felt like inside the white coat.

But Mrs. Kaufman surprised me. She pointed a spindly finger between her pale, varicosed legs and said, "The itch." She mimed scratching. "I think it's from my new boyfriend's beard."

I was silent for a minute. Her new boyfriend's beard? Then a mental image filled my head, and I officially lost it. All artifices of professionalism flew out the door as I hiccupped laughter, tears rolling down my cheeks. To my relief, Mrs. Kaufman laughed, too.

When we settled down, Mrs. Kaufman whispered, "Well, honey, if you can't tell your gynecologist, who the hell can you tell?"

Mrs. Kaufman taught me a valuable lesson that day, and I'll never forget her. Just because we age doesn't mean we go out of business. When my patients experience menopause, most feel young at heart but may need help getting their bodies to cooperate with what their young spirits desire. Whenever a woman going through menopause asks me whether she'll be able to keep her sex life alive, I tell her about Mrs. Kaufman, who continues to inspire me. I hope I can be like her when I grow up.

*I just hit menopause and I have the incredible
shrinking clitoris. I told my doctor, and she sort of
just blew it off and said it was just my age. Needless
to say, I was stunned! Can I reverse it? Is it just
going to disappear? Does this mean my days
of the big O are gone forever? Say it isn't so!*

Oh, honey. No. Don't despair. Your clitoris won't disappear, but it does tend to shrink as we get older. Because genital tissues are sensitive to estrogen, your labia, vagina, and clitoris may get smaller as you age. As estrogen levels drop, blood flow decreases. Because female genitalia, much like male genitalia, depends on blood flow, you may notice that you become smaller, paler, and flatter. Bummer, yes. But that doesn't mean your sex life is over after menopause. In fact, some women experience the best sex of their lives after menopause.

Clitoral size does not necessarily determine orgasmic strength. But if you do notice that your sexual satisfaction is diminishing, talk to your gynecologist. Your doctor can talk to you about the risks and benefits of taking hormones, such as estrogen, progesterone, and perhaps testosterone, to strengthen your tissues and improve your sex life. Most important, don't worry. The big O is yours for the taking. You just might need to give your clitoris a little TLC.

My friends who are menopausal have told me you have to "use it or lose it" when it comes to sex. Is that true? Might I "lose it" if I go through a dry spell?

It depends how long your dry spell lasts. It's true that the vagina stays healthier when you're using it with some regularity. Not only does sex keep the sensitive vaginal tissue healthy, but it's almost as if your yoni has a memory. If you keep reminding her that she has a purpose beyond reproduction, she's likely to rise to the occasion.

But failing to do so doesn't necessarily mean you'll have problems. The coochie is not like a well that dries up if you quit dipping the bucket inside. However, if you neglect your vagina for too long, the vaginal walls, which tend to become fragile and raw once there's no estrogen, may scar together and close off parts of the vagina. This can be avoided by using estrogen locally in the vagina, in the form of a cream, a pill, or a ring. Using the vagina also helps to keep it stretched out and dilated so it can accommodate your partner. If you suspect you'll have a long dry spell but hope to keep your vagina in tip-top shape, you may want to invest in a battery-operated boyfriend.

As I get older, my libido just isn't what it used to be. Why is that?

It's unclear whether declining libido happens because of beliefs or because of menopause. If your ingrained beliefs tell you that older women aren't sexual, you'll likely manifest your beliefs.

On the other hand, if you believe that women are like fine wine, improving with age, your sex life will probably reflect this.

While a large part of your sex drive lies in your mind, hormones also play a role after menopause. Falling levels of sex hormones, especially androgens like testosterone, may affect how jazzed up you feel when it comes to sex. After menopause, your ovaries continue to produce some testosterone, the hormone most associated with your sex drive. But the levels drop off, especially if you have had your ovaries surgically removed, which may reflect how you feel about getting it on. Some evidence shows that replacing testosterone to increase blood levels helps women after menopause,[1] so you might ask your doctor about whether testosterone replacement could benefit you.

What about sexual satisfaction? If you can get the desire up, is there any reason a dried-up, droopy vagina means you can't reach the same heights? It took me until perimenopause to reach my sexual prime. I'd hate to think I'm just starting to flame out....

Anything is possible. While hormonal changes and alterations in blood flow may affect some aspects of your sexuality, hormones play a relatively small role in the total picture of sexual satisfaction. You may find that you have the best sex of your life after menopause. In her book *The Wisdom of Menopause*, Dr. Christiane Northrup lists the following midlife changes you may experience:

- **Increased sexual desire**
- **Change in sexual orientation**
- **Decreased sexual activity**
- **Vaginal dryness and loss of vaginal elasticity**
- **Pain or burning with intercourse**
- **Decreased clitoral sensitivity**
- **Increased clitoral sensitivity**
- **Decreased responsiveness**
- **Increased responsiveness**
- **Fewer orgasms, decreased depth of orgasm**
- **Increase in orgasms, sexual awakening**

She says, "Reading through the list, you can quickly appreciate that change itself—and not the nature of the change—is the one common theme."

So will you flame out after menopause? Not if the flame within you burns bright. You may need to nurture the fire with a little more kindling, but there's no reason to believe you won't be able to fan the embers into a raging blaze.

What menopause symptoms make you consider prescribing hormone treatment? I'm going through menopause and my doctor didn't even mention it to me. Do I have to beg?

No, honey. You definitely shouldn't have to beg. When my patients are going through menopause, I always bring up the issue of hormones. It's a complex topic, so it takes a whole visit just to discuss the skinny on hormones. If you broach the subject at the end of your annual exam, your doctor may be too busy to properly tackle this gargantuan issue—and you deserve the best. Make a separate appointment and bring all your concerns to the table.

Whether or not I prescribe hormones has more to do with meeting my patients' needs and desires than with any particular set of symptoms. The way I see it, my job is simply to spell out the risks and benefits and let my patient choose. But certain signs and symptoms definitely trigger a discussion. Hot flashes, night sweats, mood swings, insomnia, vaginal dryness, sexual dysfunction, and osteoporosis are some of the biggies. But if you're going through menopause and have concerns, be assertive and bring it up with your doctor. We're here to help.

Am I nuts if I decide to take hormones, despite all the risks?

No. You're definitely not nuts. In fact, when the time comes, unless scary data that is more compelling emerges, I'm fully prepared to take my chances with hormones if menopausal symptoms are robbing me of my vitality. Yes, hormone replacement has its risks, but so does driving a car. Every time you get in your car, you assume the risk of getting injured or killed in an automobile accident. Yet, you still do it. Why? Because it improves your quality of life.

I think of hormones the same way. Sure, if you don't need them, why incur the potential risk? But if you live life more fully because of them, more power to you. Frankly, even if I knew hormones might shorten my life expectancy a tad, I might take that risk in the name of living a big, juicy, vibrant, passionate, radiant life. But that's just me. We all have to assess our own individual risk tolerance.

I could go on and on about hormones—why to take them, why not to—but I'll leave that to you and your doctor. Suffice it to say that when I counsel patients about hormones, I educate them fully about the risks and benefits and then support them in making their own autonomous choice. Each of us must make this difficult choice individually, with our eyes wide open.

Why did my doctor make me get off hormones? I felt great and have wanted to get back on ever since, but now I'm scared.

Prior to the Women's Health Initiative (WHI) mega-study, we believed that hormones were the Band-Aid for all that ails you in menopause—the venerable "fountain of youth." Back when I was in medical school, we were convincing ninety-year old ladies with heart disease who had never been on hormones to start taking PREMPRO. Preliminary studies suggested that hormones reduced the risk of heart disease, osteoporosis, colon cancer, even Alzheimer's. They also reduced menopausal symptoms, made your skin pretty, and kept your sex life healthy.

But whoa ... not so fast. We docs had to do an about-face, as we have many moments in medical history, when the WHI was halted. Studies suggested higher rates of heart disease, breast cancer, stroke, blood clots, dementia, and, most recently, lung cancer in women taking synthetic hormones. But these studies are far from perfect, and many of our questions remain unanswered. If you are confused about hormone replacement, you are certainly not alone. Even we docs are still scratching our heads.

Your doctor may have taken you off hormones because you have other risk factors that when compounded with the risks of hormones may put your health in danger. If you're interested in getting back on hormones, talk to your doctor or seek a second opinion from a doctor skilled in individualized hormone therapy. Ask your doctor to carefully spell out the risks and how they apply to you. A good doctor can help you understand the risks and benefits so you can make an informed decision.

Why all the hype around bioidentical hormones? Are they safer than synthetic hormones? If I take them, will I look as good as Suzanne Somers?

Bioidentical hormones are hormones synthesized in a lab to be identical in molecular structure to the hormones produced in your body. Synthetic hormones, such as PREMARIN and PREM-PRO, are intentionally created to be unnatural, since pharmaceutical companies cannot patent a natural bioidentical hormone but can patent a drug with a unique formulation. Bioidentical hormones, on the other hand, are usually made in a compounding pharmacy (although a few pharmaceutical brands, like Vivelle and PROMETRIUM, are also bioidentical hormones).

Before the WHI first released its surprising data about hormone replacement therapy in 2002,[2] most women taking hormones took synthetic hormones. The WHI studied PREMARIN and PREMPRO, synthetic hormones manufactured from *pregnant mare urine*. After results were published, interest in bioidentical hormones skyrocketed. If synthetic hormones posed such risk, maybe bioidentical hormones would be safer. Many women discontinued their PREMPRO and switched over to bioidentical hormones.

Are bioidentical hormones better than synthetic hormones? I think so. One size definitely does not fit all when it comes to hormones. Because we can compound hormones, we can customize therapy and tailor your needs to just the right hormones at just the right doses. Plus, in my experience, women experience fewer side effects on bioidentical hormones than on synthetic ones.

But are they *safer*? We honestly can't say. Because bioidentical hormone formulations cannot be patented, there is no incentive for any drug company to spend millions of dollars funding large studies to assess safety. So we must proceed with caution. While it's reasonable to believe that bioidentical hormones formulated to mimic the hormones made in our own bodies may be safer, we don't have good proof to support this hypothesis.

When my patients choose to replace their hormones after menopause, I use bioidentical hormones at the lowest possible doses to achieve our goals. But I always inform my patients that we cannot yet prove that bioidentical hormones are any safer than synthetic ones.

As for whether you'll look like Suzanne Somers, I can't say. She is certainly a beautiful woman who has clearly been blessed with some serious beauty genes. And yes, she is quite vocal about her use of high doses of bioidentical hormones. Maybe she should wear a disclaimer that says something like: "Actual results may vary."

Do old ladies have saggy vaginas?

Sometimes. Vaginas, like boobs, necks, knees, and tushies, can get saggy. Age, the trauma of pregnancy and childbirth, hormonal changes, genetics, and years of gravity can weaken the supports of the female genital tract and cause everything to sag. Not uncommonly, women whose vaginas have been well supported for decades develop pelvic prolapse when their estrogen levels drop after menopause. Estrogen is the ultimate Band-Aid for girly bits. It fluffs up the tissue, attracts healthy blood flow, and thickens genital supports. When estrogen disappears after menopause, it's kind of like letting the air out of a balloon. The balloon still exists, but it may transform into something, well, saggy.

What can I do to prevent getting a droopy box?

Don't have children. Okay, just kidding. As a mother, I can honestly say that giving birth is worth every droop. But it is also the biggest risk factor. Here are some tips to optimize your chances of keeping all your girl parts up inside you where they belong. However, keep in mind that if your DNA cursed you with piss-poor protoplasm, you might still wind up sagging.

1. **START DOING KEGEL EXERCISES NOW.** You've probably heard this before but didn't understand why. Kegel exercises are the single best thing you can do to keep your cooch in tip-top shape.

2. **MAINTAIN A NORMAL WEIGHT.** If you're overweight, gravity pushes on your pelvic parts even more.

3. **CONTROL COUGHING.** If you have bronchitis, make sure you treat your cough. Repetitive bouts of coughing cause you to do a *Valsalva maneuver* (pushing down like you have to poo), which makes prolapse worse.

4. **AVOID CONSTIPATION**. Bearing down to push out hard stool can further weaken tissue.

5. **STOP SMOKING**. Smoking makes you cough, reduces blood flow, and weakens tissue.

If you do all this and still get pelvic prolapse, rest assured that it can be treated. And yes, droopy boxes sound scary, but don't flip out. Pelvic prolapse is merely one of the small prices some of us have to pay for the joys of being women. Trust me, it's worth it. At least we don't have to worry about getting erections when we're wearing Speedos at the public pool or changing our sheets after a wet dream. As women, we get to relish sisterhood, feel a baby move inside us, speak a rich emotional language, and wear pink hats. The thought of saggy vaginas might tempt you to curse your womanhood, but don't indulge the temptation. We all must be proud of who were are, droopy boxes or not.

CHAPTER 12

Boobs

I HIT PUBERTY IN THE days of Judy Blume's *Are You There God? It's Me, Margaret.* Still completely flat-chested, I pumped my flexed arms back and forth, singing like Margaret, "I must. I must. I must increase my bust." But nothing happened.

Then one day, in the shower, I discovered a lump under my right nipple while I was soaping up my chest. I ran to my mother, crying "Mom! There's a roach in my chest!" I had an irrational fear of roaches, and growing up in swampy, sticky Florida didn't help alleviate my fears.

My mother passed this one on to my physician father, who examined my pancake-flat chest and felt the lump I showed him.

"Breast buds," he announced.

"Breast buds?" I asked.

"Perfectly natural. Just means you're starting to develop."

I heard my father go back to his ham radios, where he was busy gabbing with his buddies in South America. He said, "Sorry, guys. My daughter just had a puberty moment."

Those breast buds began a decade of breast obsession for me, ten years of peering into mirrors, crying in Victoria's Secret dressing rooms because the A cups were still too big, and fantasizing about how breast implants might make it possible to fit into a bikini without having to buy two bikini sets—a size 6 for the top and a size 10 for the bottom.

When I was a medical student, the head of the department of plastic surgery offered me a free boob job and I found myself tempted to accept.

When I mentioned it to my then-husband, he said, "Awesome!"

I said, "Wrong answer," and that's when I knew I would have to learn to live in my own skin, even if it meant never quite fitting into a bikini.

I'm not alone in my breast obsession. I had one patient, Claire, who yanked off her shirt and bra every time she was in my exam room.

"What do you think?" she'd ask, pointing to her tits.

I would repeatedly tell her that I'd perform a breast exam momentarily.

"No, Doc. What do you think? Do they look good to you?"

"Your breasts are beautiful, Claire, just like the rest of you."

She would stand up in front of the full-length mirror in the exam room, turning, holding her hands under her breasts, and looking sheepdoggishly sad. I always felt tempted to hug her. I often wondered what her home life was like. Did her husband

never tell her she looked beautiful? Why did she need her gynecologist to affirm her lovely figure? I guess we all feel uncomfortable within our own skin from time to time.

It's no wonder we obsess about boobs. Boobs in bikinis on
the beach. Boobs in beer commercials. Boobs in movies, on television, on billboards and buses, and in magazines. Boobs peek-
ing out of the *Playboy* behind the counter at the 7-Eleven. If you
were an alien assessing our society for what we care about,
breasts would be way up there.

My own fascination with the breast obsession in our society
inspired me to create an art series I call *The Woman Inside Project*. For this project, I cast with plaster gauze bandages the tor-
sos of women with breast cancer. When I complete the casting
session with each woman, I hold up the cast and say, "This is
what the world sees of you. Now tell me about the rest." I listen
for as long as it takes each woman to unveil the breathtaking
story of the woman inside. Then I transcribe the stories into first-
person narratives of the beauty I see within each woman. And
believe me, each woman *is* beautiful, regardless of what has happened to her breasts.

The same holds true for all of us. It's easy to get caught up in
how our figures look when we stand naked in front of the mirror.
But we must remember that we are more than how we look. We
are not boobs who happen to have hearts, minds, and souls; rather,
we are souls that just happen to have mounds of flesh protruding from our chests.

Looks aside, boobs can be confusing. Do we feel lumps, or is
that tapioca-pudding texture normal? Will implants or nipple
piercing keep us from breast-feeding? And are those hairs
around the nipples really supposed to be there? If you're wondering what's up with your breasts, you're in good company.

Here's to helping you and your boobs become best buds (not to be confused with "breast buds").

Why do boobs sag as we get older?

It's been a while since I've looked in a mirror, and when I got out of the tub last night I caught a glimpse of myself. Although my boobs still look perky in my mind's eye, the mirror betrays a different truth. Things just aren't the same after pregnancy, childbirth, and breast-feeding. But truth be told, I can't even blame my daughter. The girls were already starting to sag before I conceived.

Why do boobs sag? Because of gravity, plain and simple. The bigger the breasts, the deeper the droop. When we are young, our breasts primarily consist of dense glandular tissue, but as we age, fat replaces the glands, making boobs more likely to sag. Those who have given birth experience saggy breasts more often because breasts increase in size during pregnancy and as a result of nursing. Then, when milk-producing mechanisms shut off, breast tissue shrinks and what remains may resemble empty bags. Even for women who have never had children, when you hit menopause and your body knows the milk factory is officially closed breasts tend to sag further.

But don't let your spirits sink along with your boobs, ladies. This happens to *every* woman. Though it's not evident in the plastic breasts we see on television or in magazines, it is one of the many mysterious and inevitable parts of being a woman. Like us, our boobs are dynamic, shifting forms. Accept that saggy breasts are simply the price we pay for the inner beauty we gain with the wisdom of age.

I hate bras. If I don't wear one, will my boobs droop?

I hate bras, too. When I'm home writing, I shun them, but like most of us, I conform to what's expected when I'm out in public. Sure, wearing a supportive bra will keep your boobs from drooping—until you take it off. The minute you remove the bra, the sag sets in. Contrary to what my mother insists—that her perky breasts are due to the fact that she has worn sturdy support bras her entire adult life—wearing a bra does not seem to prevent breasts from sagging over the long haul. Because breasts consist of glands, connective tissue, fat, and ligaments, there is no muscle to tone by wearing a bra. The degree of breast sagging each women experiences (called "ptosis") depends more on your genes, your diet, your hormones, your breast size, and the elasticity of your skin. In fact, several studies suggest that wearing a bra may even increase breast sagging.[1]

Susan M. Love, M.D., author of *Dr. Susan Love's Breast Book*, writes: "A mistaken popular belief maintains that wearing a bra strengthens your breasts and prevents sagging. But you sag because of the proportion of fat and tissue in your breasts, and no bra changes that.... If you enjoy a bra for aesthetic, sexual or comfort reasons, by all means wear one. If you don't enjoy it, and job or social pressures don't force you into it, don't bother. Medically, it's all the same."[2]

So relax and let it all hang out. Wear a bra because it's comfortable, or don't wear one if you feel better that way. Going commando will probably have no impact whatsoever on boob droopage.

My boobs are two totally different sizes. I call them "Flip and Flop, a tale of two titties." Is that normal?

Body parts are often asymmetric. My left foot is a size 7, while my right foot is almost an 8. The rings that fit my right hand are too big for my left hand. We don't think much of it if our feet or fingers are different sizes. Boobs are the same way. No two are exactly alike.

It's quite common to have some asymmetry in breast size. My patient Serena called herself the Cyclops because one breast was a perky A cup and the other was a slightly droopy C cup. She always felt self-conscious and considered getting an implant on one side until her boyfriend confessed that he loved her lopsided breasts. Think of your asymmetric breasts like you would of a crooked smile or a beauty mark—they're all subtle imperfections that make you all the more endearing and unique.

I try to do breast self-exams, but my boobs always feel lumpy. How can I examine my breasts without freaking out and running to the doctor every time I do it?

Self-breast exams can be nerve-wracking if you're not sure what you're doing. You're lying in bed at night, feeling your boobies, when all of a sudden you feel something lumpy. It's ten o'clock at night, you're all alone, and now you're thinking you're dying of breast cancer. It's enough to drive you bonkers.

So what's a girl to do? Trying to find lumps in something naturally lumpy-bumpy can be tricky, which is why I recommend that my patients examine their breasts right after I examine them. Next time you see your doctor, ask for a breast exam. When your doctor tells you your boobs feel normal, go home and learn what normal feels like for you. Every breast is different. Yours may feel like tapioca pudding or cottage cheese, with little bumps all over. Or they may feel very soft and fatty, like your butt cheek (or I should clarify, like *my* butt cheek). Your boobs may feel almost muscular, like the back of your calf. How they feel depends on how glandular they are and whether or not you have a normal condition called "fibrocystic change," which can make boobs bumpier.

Try to feel for something you can pick up and move around, kind of like a small rock or a marble. Most important, you're feeling for something new. While we docs tend to be experienced in knowing what to feel, you've got a leg up on us because we only examine your breasts once a year. You can get up close and personal with them every day if you wish. If you know your breasts well, you're more likely to discover a lump if it appears.

While new recommendations from the U.S. Preventive Services Task Force recommend against self breast examination, I can't see the harm in learning to be aware of changes in your own breasts. I've known many women who find their own breast cancers, and personally, I'm going to keep right on checking. If you examine your breasts and something seems amiss, don't panic, but do call your doctor.

My nipples stick inward. Will I still be able to nurse my babies? Could it be a sign of breast cancer?

Just like belly buttons can be innies or outies, nipples can protrude outward or retract inward. In fact, 10 percent of women have inverted nipples, and most of the time it's perfectly normal. You may have heard that nipple inversion can be a sign of breast cancer, and that's true. If you've always had outies and now all of a sudden one nipple is inverted, see your doctor. But if you've always had inverted nipples, that's probably just the way you were made.

Most women with inverted nipples can successfully breast-feed. Some babies will struggle to latch properly if your nipples are inverted, so if you're having trouble getting your baby to nurse, ask to see a lactation consultant. They have tricks up their sleeves that can encourage your baby to suckle.

If I pierce my nipples, will they still work for breast-feeding?

All you women out there with nipple rings (or those considering getting them) can breathe a sigh of relief. You can be a hip, bejeweled, pierced mama and still breast-feed. A normal nipple consists of a whole network of tiny milk ducts, so it's not just one big milk spout. Even if your piercing interrupts a few milk ducts, the others should work just fine. Unless your nipple piercing gets infected and scars during healing, your baby will likely guzzle away at the milk bar without a problem.

However, you may need to remove your nipple jewelry to nurse safely and effectively. Lactation consultants at La Leche League recommend removing nipple jewelry in order to avoid potential complications. While some women can nurse successfully with jewelry in place, this may lead to problems with improper latching and babies who have trouble with gagging, slurping, or leaking milk out of their mouths. Also, your baby could choke on the jewelry if—God forbid—it should become un-fastened.

What happens when you're done nursing? Will you need to get repierced? Elayne Angel, author of the *The Piercing Bible: The Definitive Guide to Safe Body Piercing*, says, "If you decide to take out your jewelry and leave it out until you are done nursing, the piercing may shrink or close up by the time the baby is weaned. If your piercing is fully healed, there is some chance the hole could remain open. It may be possible to encourage a well-established channel to stay viable by passing a small, clean insertion taper through it on a regular basis. If the piercing has sealed shut and you wish to be repierced, it is best to wait at least three months after you stop nursing to allow the tissue to normalize."

If I have breast implants, can I still breast-feed?

Probably. Most of the time, women who have had cosmetic breast implants can successfully nurse their babies. But sometimes babies have to settle for bottle-feeding. Whether or not you can breast-feed depends largely on how your implants were inserted. If they were inserted through incisions under the fold of the breasts or under your arm, chances are good that you will breast-feed successfully, because these incisions do not interrupt the nerves or the milk ducts in the nipple. If your implants were inserted through a "smile" incision around the areola of the nipple, your risk of having trouble with breast-feeding is higher. If your areola was incised and the milk ducts and nerves were not severed, it's likely your infant will be able to slurp away happily. But if the nerves that trigger the brain to release the hormones prolactin and oxytocin were damaged, or if too many of the milk ducts were severed, you may be unable to breast-feed.

How do you know if you'll be able to nurse? You don't. You'll just have to try and see. There's no way to predict beforehand. If you're producing some milk but not enough to fully nourish your baby, don't be discouraged. Even if you have to supplement with formula, keep feeding your baby what you do produce. Every little bit of breast milk counts.

If you're considering getting breast implants but want to be sure you can breast-feed, wait to get your implants until after you've nursed your last child. Then you can pump up your knockers without worry.

My boobs are so big my back aches and people stare. Should I get a breast reduction?

The grass is always greener, isn't it? While some of us would die to fully fill out a B cup, others suffer from carting around a rack of double Es. If you're one of those women with massive breasts, should you get a breast reduction? That's a decision only you can make. In my experience, women choose to reduce the size of their breasts for very different reasons than women who augment them. We small-breasted women have no idea how handicapped some women with enormous breasts feel.

My patient Elise, who is thin, pretty, and has humungous hooters, said, "Can you imagine never being looked in the eye by a guy who is talking to you? The Tits walk into a room a full minute before I do, and by the time I arrive the Tits have attracted all the attention. The rest of me doesn't stand a chance. And don't even get me started about that marathon I'm training for. Believe me, no sports bra will rein those babies in. The pain in my bouncing breasts and the backaches that plague me daily make me curse the body I was born with."

My heart goes out to women like Elise. While I'm not a big fan of plastic surgery, I have written many letters to insurance companies on behalf of my patients with big boobs. Believe me, nobody undergoes this surgery for purely cosmetic reasons. Breast reduction surgery is painful, disfiguring, requires a long recovery, and may affect the ability to breast-feed. But I have yet to meet a woman who regretted it.

Is it normal to have hair on my nipples?

Yes. Many women have hair on their nipples, which can range from thin and light to coarse and dark. Whether or not you have hair on your nipples tends to reflect your genetic makeup and your levels of androgens (male sex hormones like testosterone that normally exist in lower levels in females). If your mother tends to bear certain resemblances to the gorilla at your local zoo, or if you have higher levels of androgens coursing through your body, you're more likely to wind up with course, dark hairs on your nipples. Chances are, if you have hairy nipples, you may also have hair on your upper lip or chin.

How do you know the difference between normal hairy nipples and the hair being a sign that something's wrong? Pay attention to the distribution of the rest of your hair. Do you have dark hairs in the center of your chest, where guys get it? Or dark hair coursing from your belly button down to your mons pubis? Or excessive facial hair? If so, tell your doctor. This could signal abnormally high levels of androgens, which may need to be evaluated.

Why do boobs get sore before your period?

During the second half of the menstrual cycle, boobs have a tendency to swell from fluid retention, which makes them tender. Breasts may also become more tender in perimenopause as the hormones fluctuate. Caffeine, which contains methylxanthine, a chemical that can cause blood vessels to dilate, may exacerbate this type of breast tenderness.

So I hate to break it to you, but you might feel better if you skip that grande caramel latte with the extra shot of go juice. A diet high in sodium also leads to breast swelling and tenderness. Cutting back on dairy products and animal fats may reduce boob soreness, as well. So try modifying your diet if cyclic breast pain plagues you.

If this doesn't help, talk to your doctor about getting your hormones checked. Women with higher levels of estrogen may experience more breast tenderness and may find some benefit from treatment with progesterone.

Is it normal for breasts to hurt when I don't have my period? Why does that happen?

We're trained to associate pain with disease, so when women experience breast pain they often wind up in my office wigging out because they think they have breast cancer. If you're one of these women, take a deep breath. Most breast cancers are painless, and most breast pain (we call it "mastalgia") has nothing to do with cancer. In one study, 45 percent of women experienced breast pain and 21 percent called their breast pain "severe."[3]

So why do breasts hurt? Breast pain can be either cyclic (usually occurring premenstrually) or noncyclic (boobs hurt all the time).

Noncyclic breast pain is more likely to be related to other factors, such as large breasts, hormonal contraceptives, hormone replacement therapy (HRT), dietary factors, or other rare breast conditions. Some believe stress can trigger mastalgia as well. If you're worried about your breast pain, see your doctor to make sure everything is A-OK, but don't stress out—that just might make it worse.

I used to love having my breasts touched, but after nursing my son I feel nothing. Can breast-feeding decrease sensitivity in your nipples?

While I can think of no scientific mechanism to explain this, many of my patients complain about the same thing. After nursing two boys, my patient Jillian pointed to her boobs and said, "The way I feel, those things belong to Jordan and James [her sons]. John [her husband] doesn't stand a chance."

While breast-feeding shouldn't damage the nerves or cause any permanent changes that affect arousal, I've heard this complaint too often to dismiss it. When I informally surveyed my patients, results were split. Some said breast-feeding made their nipples more sensitive, some reported no change, and a disgruntled percentage swore that nursing ruined their ability to be aroused by nipple stimulation.

Because sexual arousal for women arises as much from our minds as from our bodies, I question how much this issue springs from how we relate to our breasts. Perhaps, once we use our breasts for biological, rather than sexual, purposes, we quit thinking of them as opportunities for sexual arousal. Maybe a little ceremonial breast rebirth is in order. By reclaiming your breasts for your own pleasure and releasing them from the valuable job they served as baby feeders, perhaps you can once again feel sexually aroused when your lover stimulates your breasts.

I have three nipples. Why am I such a mutant? How common is it to have two nipples on one side?

If you have three nipples, you are in good company. This condition exists in 2.5 percent of humans and is nothing to worry about.[4] This variation occurs when an extra nipple grows along the milk line when you're still in the womb. Milk lines are two invisible paths that start at the armpits and traverse the normal nipples, ending at the groin. While many mammals, such as dogs and pigs, have nipples all along the milk line, humans usually have only two. During embryonic development, a mammary ridge develops up and down the front of the torso, but most of the time only two nipples remain.

Sometimes, things go awry in fetal development and you wind up with an extra nipple or two, which we call "supernumerary nipples." They may be mistaken for moles until a woman gets pregnant or starts nursing, when all of a sudden the nipple grows and may lactate. (Zoinks, Shaggy!) If you have a supernumerary nipple, you are not a mutant. Instead, you might consider yourself breastfully blessed.

After I had a baby, my nipples got big and brown.
Will they ever go back to being small and pink?

When you are pregnant, high doses of estrogen course through your body. Because estrogen can affect skin pigmentation, you may notice certain parts of your body darkening during pregnancy, including your nipples. These changes may persist once your pregnancy ends, especially in women with darker complexions. In addition to causing your nipples to grow and darken, estrogen can cause *melasma* ("the mask of pregnancy"), which results in dark splotches on your cheeks, forehead, or upper lip. You also may notice a dark vertical line called the "linea nigra," which runs up and down your belly, and birthmarks may darken under estrogen's influence. Usually, these changes fade over time, as estrogen levels return to normal. But these changes can sometimes be permanent.

I quit nursing a year ago, but when I squeeze my
nipples milk still comes out. Why is that?

Most of the time, breast milk dries up within a year. If you're still producing drops of milk when you squeeze your nipples, chances are you're squeezing your nipples too much. Every time you stimulate those nipples, you trigger hormones that tell your brain (and your breasts) that a baby wants to eat. To convince your body that you're serious about weaning, avoid all nipple stimulation for a while. Break it to your partner that your nipples are off-limits during foreplay, and keep your own hands off your boobies. Avoid wearing clothing that rubs your nipples or holding your baby in a front carrier. If you give your nipples a break from being touched, they will realize that there's no baby to feed and your milk should dry up completely. If it doesn't, see your doctor. Rarely, tumors in the brain can release prolactin, leading you to lactate when you're not nursing.

My nipples are the same color as my skin. Is that normal? Is there anything I can do to make them darker?

Usually the nipple and areola are darker than the rest of your skin, but this isn't always the case. Some women have nipples the same color as their skin, while others may have nipples slightly lighter than their skin. All are variations of normal and nothing to worry about. If this makes you self-conscious and you wish to change the color of your nipples, you can tattoo your nipples the color you choose. But do you really want to go through that?

Is it true that the Pill can make your boobs grow?

Good news for small-breasted woman everywhere! Because breasts are estrogen sensitive and the Pill contains estrogen, yes, the Pill can make your boobs grow. Many women notice that their breasts increase a whole cup size when they're on birth control pills (the same is true for the hormonal birth control patch or the vaginal ring). Sounds like cause for celebration, doesn't it? A visit to the gynecologist for a birth control prescription, and then off to Victoria's Secret to buy some new lingerie!

Not so fast, ladies. This is not some miracle alternative to breast implants. While many do notice an increase in their bust size, some do not, and the effect is not permanent. As soon as you stop the Pill, the girls usually go back to their original size unless you have put on weight. Keep in mind that the Pill is not without its risks, so I don't recommend using it solely for breast enhancement.

Do most breasts increase, decrease, or stay the same size after breast-feeding?

I'd have to choose "all of the above." While breasts uniformly increase in size during pregnancy and lactation, how breasts go back to "normal" varies widely. Some women's breasts increase in size after pregnancy, especially if they have been unable to lose the baby weight. Because boobs consist mostly of fat, any additional weight may wind up on your chest. Some women notice that when all is said and done their breasts shrink to a smaller size than before they got pregnant. But most notice no change in cup size. So while the shape and appearance of breasts is likely to change as the result of pregnancy and breast-feeding, the majority of women fit back into their old bras once they get their post-baby body back.

Do bigger boobs make more milk?

Nope. Even if you're flat as a pancake, you should be completely capable of being the milk goddess your baby needs. All female breasts are equipped to be full-service milk-making machines. The only difference between large breasts and small breasts is fat. Regardless of size, each breast contains fifteen to twenty-plus milk-producing mammary lobes, which should produce as much milk as your baby needs. In fact, lactation consultants say that small-breasted women may have an advantage, since babies may find it easier to latch onto a smaller breast.

How long should I breast-feed?

I recommend that all mothers try to nurse for a bare-bones minimum of six weeks and up to a full year if they possibly can. If you're blessed with time, patience, cooperative breasts, and a willing baby who makes this possible, you're blessed. And if it suits you, nursing for more than a year rocks. Every little bit helps, so even if you have to supplement, try to squeeze out a wee bit of breast milk to share with your baby. But give yourself a break if you've done everything possible and it's just not working for you. Some women do everything right and still can't nurse, so don't beat yourself up about it.

Why do they have to flatten your tits between two plates to do a mammogram? Clearly, some guy invented that horrible machine. Is it absolutely necessary that I do them, and if so, when should I start and how often should I get them?

My father was a radiology doctor who specialized in mammograms, so I spent a lot of time growing up around those machines that take two perfectly good knockers and squash them into Swedish pancakes. Dad used to joke that if women were in charge of medicine, men would have to stick their penises between two plates and get "manograms" after the age of forty.

When I was young, I remember sitting beside Dad in the dark screening rooms while he inspected film after film, hunting for subtle signs that might signal breast cancer. He likened it to reading a *Where's Waldo?* book. Mammograms may seem like torture, but they may save your life.

While mammograms are far from perfect as a screening tool for breast cancer, studies estimate that they detect about 75 percent of breast cancers in women in their forties and 90 percent percent of breast cancers in women in their fifties and sixties.[5] Early detection can mean the difference between living and dying. If you're over forty and you get mammograms regularly, your risk of dying of breast cancer is reduced by 34 percent.[6]

Nobody in the medical community seems to agree on the best way to offer breast cancer screening. All major medical societies recommend mammography after the age of fifty, but what to do before then and how often to screen is controversial.

Personally, I recommend that my patients get mammograms at least every other year after forty. I know that mammography has its limitations and I pray that we will find a safer, more effective, more humane screening tool soon, but in the interim, I know way too many people who were diagnosed with breast cancer in their forties, and if there's anything we can do to promote early diagnosis, I say let's do it. But stay tuned. Recommendations are changing. Talk to your doctor to keep up-to-date.

I just turned forty, so I'll soon be making the phone call to get my hooters mashed between two plates. The way I see it, I've got a lot to live for, and I'll do whatever it takes to keep my boobies healthy.

I'm terrified of breast cancer, and even though I'm only thirty, I want to start getting mammograms. But my doctor recommends against it. Why is that?

I don't blame you for being confused. If mammograms reduce your risk of dying from breast cancer, why don't we start doing them from the time we sprout breast buds at puberty? Unfortunately, it's not so simple.

Although mammography clearly benefits older women, the data is fuzzier when it comes to young women. Because younger women have perky, dense, glandular breasts, X-rays don't penetrate the breast as well. Although data suggests that mammograms benefit women after the age of forty, they don't work so well for younger women. Nevertheless, one study found a 15 percent reduction in breast cancer deaths when younger women were screened.[7] If young women have a first-degree family member (mom or sister—not grandmother or aunt) who was diagnosed with breast cancer at a young age, I recommend that they start getting mammograms five years younger than the family member got cancer. So if Mom got breast cancer at thirty-eight, you'd start at thirty-three.

Screening in younger women is not without its costs. You can expect a higher risk of false positives, a higher rate of unnecessary procedures, a great deal of expense, more stress,[8] and some women feel concerned about exposing their delicate breast tissue to radiation. One study estimated that you would have to screen 1,792 younger women to prevent 1 woman from dying of breast cancer.[9] Another showed that you would have to do ten thousand mammograms to prevent one death each year.[10] But if you're that one woman, you're probably happy you did it.

Breast thermography provides an alternative to mammography in young women (as well as a good adjunct to mammography) and may be particularly useful in young women with a strong family history. Breast thermography, a digital infrared imaging technique that assesses heat in the breasts, may reflect the increase in blood vessels associated with evolving cancer. Breast MRI provides another alternative, though it's an expensive procedure and your insurance carrier is unlikely to cover it. But talk to your doctor. Things are changing quickly, and you want to make sure you're on the cusp of the latest breast cancer research.

I'm terrified of getting my boobs lopped off. Is there anything I can do to prevent breast cancer?

To decrease your breast cancer risk, here are some tips.

1. Have your children at a young age.

2. Breast-feed for at least six months.

3. Minimize the use of hormones after menopause.

4. Maintain a healthy weight.

5. Exercise regularly.

6. Practice monthly self breast exams.

7. Eat five or more servings of organic fruits and vegetables per day.

8. Limit your intake of animal fats, particularly red meat.

9. Avoid alcohol, or limit it to no more than one or two drinks per day. If you do drink alcohol, take a folic acid supplement, which moderates this risk.

10. Increase your intake of superfoods high in antioxidants, such as kale, beets, carrots, beans, collard greens, brussels sprouts, and broccoli.

11. Drink green juice.

12. Avoid dairy or use organic butter, cheese, and milk, as they are less likely to be contaminated with human growth hormone or estrogen, which is sometimes used to stimulate milk production in cows.

13. Use extra-virgin olive oil, raw flaxseed oil, and cod liver oil.

14. Expose yourself to the sun, which increases your levels of vitamin D.

15. Get mammograms (or investigate alternatives).

16. Know your family history. If you have a first-degree family member who was diagnosed with breast cancer before menopause, consider talking to a genetic counselor.

17. Be aware of xenoestrogens, environmental chemicals that act like estrogen in the body and may increase your risk of breast cancer. While you can only do so much to avoid xenoestrogens in your own life (eat

organic, avoid plastic, et cetera), heightened awareness and caring for our planet may help save us all.

I hear these scary statistics about breast cancer risk. What's the real scoop?

We all hear the scary statistics—one in eight women will get breast cancer in her lifetime.[11] Which is true, but misleading. So what does it mean? Yes, the lifetime probability of developing invasive breast cancer is one in eight. But statistics can be very confusing. Because we live longer and have better screening methods, our lifetime risk of breast cancer keeps increasing. When you are born, you have a 12.3 percent chance that you will be diagnosed with breast cancer in your lifetime but less than a 3 percent chance that you will die from it. When you are twenty, you have a 1.9 percent chance that you will be diagnosed before you are fifty. At fifty, there is an 8.9 percent chance you will be diagnosed in the next thirty years.[12] So it's not as if, every year, you have a one-in-eight chance of being diagnosed with breast cancer. But it's frightening, nonetheless.

I've heard using antiperspirant can cause breast cancer. Is this true?

Surely you've all gotten the e-mails, the ones that scare the hell out of you for even thinking about slathering on antiperspirant before you head to the gym. So what's the science behind all the hoopla?

Researchers at the National Cancer Institute and the National Institutes of Health state: "There is no conclusive research linking the use of underarm antiperspirants or deodorants and the subsequent development of breast cancer." So what about all the rumors? Some research suggests that aluminum-based compounds, when applied to the skin, can be absorbed and may cause estrogen-like effects in the body.[13] Other studies blame parabens, which may be present in some deodorants and antiperspirants and may also have estrogen-like effects.[14] Some link it to a combination of armpit shaving and use of these products.

Studies to date show differing results. Two studies suggested no increase in risk,[15] while another suggests a possible link between antiperspirant/deodorant use and underarm shaving with breast cancer.[16]

What can we make of all this? The way I see it, you can't live in fear. I know some women who are so busy trying to avoid things that might cause them harm that they forget to live. Life's too short. Plus, I don't know about you, but personally, I'm going to keep on using my Secret Solid. I'm just not willing to tolerate icky, sweaty pits, but that's just me....

CHAPTER 13

Pee

WHEN I FIRST LEARNED ABOUT urinary tract infections, I was a second-year medical student, newly in lust with Kirk, a fellow medical student and the guy who eventually became my first husband. One rare weekend away from school, Kirk swept me away in his shiny red Porsche to an oceanview room in an old historic Florida resort hotel. I shopped at Victoria's Secret for fancy new lingerie, and we spent the weekend exploring each other's bodies while the waves crashed outside our balcony.

The following Monday, I was back in school, sitting in lectures from 8 A.M. to 6 P.M., reluctantly gobbling down the basic science education required to move on to clinical rotations the following year. My first day back from our weekend tryst, I crossed my legs, trying to make it to our first restroom break, darting straight to the potty as soon as the professor excused us. I peed, just a little bit, and even though I still felt like I had to go, nothing more came out. I wiped, stood up, pulled up my panties, and started to leave the stall. Right then, I was hit by another urgent need to pee. I pulled down my pants and sat on the pot, waiting for something to come out, but nothing did. Ten minutes of burning pain ensued, and when I finally eked out one or two drops, the blood-tinged droplets discolored the whole toilet bowl. Something was *not* right.

I suffered through the next few hours of lectures and, over my lunch break, I beeped my girlfriend on her pager. She was a year ahead of me in medical school and was already doing her clinical rotations (which meant she actually knew something about clinical medicine). When I described my symptoms, she said, "Oh, poor baby. You have a urinary tract infection. Did you and Kirk get it on this weekend?"

I blushed. How did she know? When I asked, she said, "That's how it happens. You bump like bunnies, and then—*boom*. Honeymoon cystitis."

"Honeymoon cystitis?"

"Yeah. You go on your honeymoon, you have lots of sex, and you wind up with a UTI. It's a well-known fact."

Well, someone could've told me that *beforehand*.

My friend hooked me up with some antibiotic samples (the upside of being a medical student—prescription drug samples are plentiful). A little cranberry juice, some vitamin C, and a dose of bladder anesthetic, and I was good as new. But I'll never forget that day. I can remember what I was wearing, which bathroom I used at the medical school, who sat next to me in class while I crossed my legs. It's almost like the day the *Challenger* exploded, or September 11. Well, I suppose it's not exactly the same, but if you've ever had a rip-roaring UTI, you know what I mean.

My urinary life normalized after that and remained a nonissue until I got pregnant, shortly after marrying my third husband. (I know. I know. It's not something I'm proud of.) We planned our honeymoon strategically, since we were going to try to conceive even before the trip. We figured a Hawaiian vacation would allow us to have fun, even if I was puking my guts out and exhausted.

As it turned out, I was sixteen weeks pregnant during our trip and felt great—so good, in fact, that we planned a series of ten-mile hikes. What I failed to consider was what it would mean to be sixteen weeks pregnant and hiking a crowded trail, miles from any bathroom. I'm not kidding when I say I had to stop every fifteen minutes to pee. Yes, I had to use banana leaves to wipe my cooch because I failed to plan ahead and bring a whole roll of toilet paper. Yes, I got busted with my bare ass sticking out. Yes, my husband got sick of explaining, "Forgive my wife. She's pregnant." Let's just say that pee problems can be a lesson in humility.

Patients and friends can attest to that. My friend Annie insists on jumping on the trampoline with her three kids, even though she's guaranteed to leak urine all over her clothes. When her littlest one saw her wet pants, she said, "Mommy, it's okay. I still have accidents, too." After that, Annie's husband bought her a box of Depends—a joke she didn't find terribly funny.

Urinary tract infections, incontinence, and other urinary mishaps may get you down, but you'll get back up again.

When I laugh or sneeze, I always leak pee. What's up with that?

You probably have stress urinary incontinence. The name might suggest that it's caused by emotional stress (it's not) or that it causes stress (it can). Actually, stress urinary incontinence is usually caused by a change in the angle of the urethra, the tube that connects the bladder to the outside world, and a weakening of the urinary tissues. Predisposing factors, such as childbirth, aging, and genetic factors, may make you more susceptible to leaking urine under "stress," which in this case entails things like laughing, coughing, or sneezing. When this happens, the pressure inside the abdomen exceeds the urethral sphincter's ability to hold in urine. And voilà—you leak. But fear not—you're not doomed to this fate for all eternity. Read on....

Have you ever known Kegel exercises to really work and eliminate incontinence?

Absolutely. In fact, my patient Elise did her Kegel exercises so religiously that her boss Andrea started suspecting something was seriously wrong with Elise. Elise would be sitting in a board meeting with a strained, serious look on her face, and when someone asked her a question she seemed clueless, as if she hadn't been paying attention. During lunch breaks, Elise seemed distant and distracted. In the bathroom, Andrea noticed that Elise spent extraordinary amounts of time in the toilet. Finally, Andrea called Elise into her office.

Andrea confessed that she and the rest of the team suspected that Elise might be drinking on the job, using drugs, or suffering from some sort of mental illness. Elise seemed so inattentive, and Andrea couldn't begin to imagine what she was doing in the bathroom for so long. They asked her to take a drug screen, which Elise passed with flying colors. With her job on the line, Elise finally confessed. She was plagued with urinary incontinence, and when I told her to do her Kegel exercises religiously she took it to heart. She did Kegels in the boardroom, Kegels in her office, Kegels in the lunchroom. And whenever she used the toilet, she practiced starting and stopping the stream of urine until all of the urine was released.

When Elise explained what was happening, Andrea asked, "Well, is it working?" Turns out that Andrea, another patient of mine, was also suffering from stress urinary incontinence but didn't believe that Kegels could actually work. She and Elise began egging each other on at work, reminding each other to do Kegel exercises (on break, instead of in the boardroom). Both reported significant improvement in their symptoms.

Elise and Andrea aren't alone. Studies demonstrate that when incontinent women are compliant with these safe, cost-effective pelvic muscle exercises, they demonstrate improvement when compared to placebo treatments.[1] And as an added bonus, your orgasms may strengthen in intensity (woo hoo!) So yes, Kegel exercises may be worth the effort. They don't always work and some patients need to pursue further treatment, but why not try? You've got nothing to lose (as long as you don't lose your job!).

Why is pee sometimes dark yellow and sometimes clear? What's better?

If your urine is dark yellow, you're probably dehydrated. Urine should be a very light straw color or, even better, clear. When you're not consuming enough liquids, your kidneys concentrate all of the toxins they filter into the small amount of liquid they can spare. Dark urine is usually your signal to drink more—lots more. Once your urine becomes clear, you're probably adequately hydrated.

Other factors that can affect the color of your urine include medications you're taking, foods you eat, and blood in the urine. Rarely, dark urine can be a sign of liver disease. If you drink lots of water and your urine is still discolored or if you don't feel well, err on the side of caution and see your doctor.

Why does asparagus make my pee stink? And why does my boyfriend say it doesn't happen to him?

Surely some of you have noticed this phenomenon. Spring comes, and you gorge on the season's yummiest veggie, only to be stopped in your tracks an hour later when a foul odor wafts up from the toilet bowl. If you didn't know better, you might suspect that you're rotting from the inside out.

If you've ever eaten asparagus and noticed a rank odor emanating from your urine, you are not alone. About half the population report smelly pee after eating asparagus. The other half deny any change in the smell of their urine. Why is this? Asparagus consists of sulfur-containing amino acids, which, when broken down during digestion, release sulfurous components that give aspara-pee its distinctive odor. While scientists debate the actual compound responsible for the smell, it clearly passes through the system quickly. My best friend and I once timed how quickly we could smell the funk after eating asparagus. Mine took twenty-three minutes. Hers was thirty-five. But then again, I ate more asparagus.

Why do only half of the population notice stinky pee after eating asparagus? Scientists are divided. Some think only half the population have the gene that breaks down asparagus into its odorous products. The other camp thinks only half of us have the gene that allows us to smell the sulfurous by-products. I tend to believe the latter. One of my girlfriends insists her pee doesn't smell after she eats asparagus, so I asked her to prove it to me. After sniffing her toilet bowl (hey—anything in the name of science!), I said, "Pee-ew!" She looked baffled. Turned out she couldn't smell it, but I could. Go figure!

If my pee smells bad on a regular basis, does this mean something might be wrong with me?

Usually, healthy urine has almost no odor (unless you've been feasting on asparagus or other foods or vitamins that are known offenders). If you are dehydrated, your urine will be more concentrated and may have a stronger smell. Also, if bacteria have contaminated the normally sterile urinary system, which happens when you have a urinary tract infection, you may notice an odor. Rarely, metabolic diseases can give the urine an unusual odor.

Most of the time, when women approach me complaining of foul-smelling urine, they're actually smelling a vaginal odor. If you have a vaginal infection, such as bacterial vaginosis or trichomonas vaginalis, you may notice that your vaginal discharge, when it mixes with your urine, smells icky.

If you're not sure what's wrong, see your doctor and let them help you sort it out.

If I think I have a bladder infection and I have some leftover antibiotics in the cabinet, can I just take them?

Please, please don't. Unless you suffer from frequent urinary tract infections (UTIs) and your doctor has specifically prescribed extra antibiotics aimed at helping your poor bladder, it's best not to go raiding the medicine cabinet. First, you want to make sure it's really a bladder infection and not something else, like a kidney infection, kidney stone, or interstitial cystitis, which likes to mimic UTIs. Plus, we doctors carefully select antibiotics to treat for the most likely bacterial types. If you use the antibiotic your doctor prescribed for your sinus infection, there's a good chance it won't treat your bladder infection, and it may land you with a rip-roaring yeast infection to boot.

Bottom line, ladies: If you think you have a UTI, call your doctor. If you suffer from frequent UTIs, ask your doctor for refills so you can get more antibiotics if you need them pronto.

Why do I get insane bladder infections after a night of wild lovemaking?

Ah, the old honeymoon cystitis—a hunka hunka burning love. Here's the deal. The bladder and urethra are supposed to be sterile. Although bacteria live on the vulva, they don't tend to climb into the bladder without some help. Anything that introduces something from the outside world into the bladder, such as a bladder catheter, puts the bladder at risk of infection. Although it's not so direct, sex acts the same way, by introducing bacteria into that fragile urethral environment. Every woman's anatomy is different, and this explains why some women suffer from UTIs every time they get freaky and others don't. For example, if the distance between the urethral opening and your bladder is short, bacteria need to travel a shorter distance to wreak havoc.

Are there ways I can reduce the risk of getting UTIs after sex?

Yes. Here are a few tips:

- Empty your bladder before and after sex.
- Make sure you wipe from front to back to avoid contamination of the urethra with rectal bacteria.
- Drink plenty of fluids.
- Take vitamin C supplements daily, which increase the acidity of the urine, making it harder for bacteria to grow.
- Take cranberry supplements.
- Wash your genital region before intercourse, and have your partner do the same.
- Try to avoid any cross-contamination between the anus and the urethra. If you engage in anal sex play, keep all contaminated body parts and sex toys far away from the urethra.
- Experiment with different sexual positions that reduce friction around the urethra.
- If nothing else helps, talk to your doctor about taking one dose of an antibiotic before you have sex each time. While regular use of antibiotics is best avoided, this may be the only way some women find relief from recurrent UTIs.

CHAPTER 14

Butts

SOCIETY GIVES US MIXED MESSAGES about butts. On one level, we're led to believe they're almost as valuable as boobs. As a kid, I grew up singing the song from *A Chorus Line* over and over again. Tits and ass, right? I'd puff up my chest and stick out my hiney, pretending I, too, was a chorus girl. The boobs never really showed up, but the ass appeared when I was quite young.

My mother always said, "You have a nice, healthy butt just like your father." *What would an unhealthy butt be like?* I wondered.

My half-African-American sister, Keli, said, "Girl, you got a black-girl ass." Apparently, it's true. As a kid, I was a scrawny white girl with knock-knees and a bubble butt.

My nurse Sandy summed it up. She said, "Honey child. Me, I'm a boobie doo. My boobies stick out more than my booty do. You, you're a booty doo—your booty sticks out more than your boobies do." Urban dictionaries describe the terms differently, but I think I prefer the way Sandy described it. I'm definitely a booty doo. And that's just fine by me. Because of that, I've always attracted ass men, which is kind of bizarre to me. I mean, what's all the fuss about? So I've got two hunks of gluteus muscle. Big whoop.

Popular music supports the notion that butts matter. Check out E.U.'s song "Da Butt," Sir Mix-A-Lot's "Baby Got Back," and Queen's "Fat Bottomed Girls." According to music, big butts rock. So why are so many of us self-conscious of our butts? And of course, songwriters are referring to the outside parts—the butt you can see through our jeans.

But what about the rest? Underneath all the curves lie the biological processes most of us would prefer to ignore. Between those butt cheeks lives the anus, the tail end of the gastrointestinal tract. From that hole, poo, farts, and other dirty, smelly unmentionables emerge. How do we resolve this paradox, the juxtaposition of the sexual with the biological?

When I told a friend I was going to try to answer all of your butt questions, she said, "Just take a crack at it." We both giggled. But that's how it goes with butts, isn't it? We make butt jokes, keep it light, and avoid talking about butts as much as possible, leaving the poor bum neglected and sorely misunderstood.

I've discovered that one of the things women really want to discuss but are too embarrassed to bring up is anal sex. Hannah was one of those women. She said, "My boyfriend is obsessed with my butt and I don't know how to handle it."

Upon further questioning, she admitted that he preferred anal sex to vaginal sex and, frankly, she was sick and tired of taking it up the butt. She cared for him, and other than their sex life she felt happy in the relationship, but she didn't understand why they couldn't just have sex like normal couples. She didn't get any pleasure out of anal sex, and sometimes it hurt. She told him how she felt, but he insisted. They tried compromising, but ultimately, it came between them and they broke up. After the breakup, Hannah wanted to know whether there was something wrong with her. Did other women hate anal sex, too, or was she alone in how she felt? I told her she definitely is not alone.

But some women genuinely enjoy anal sex. Jessica and her partner view anal sex as a special treat they reserve for romantic weekends away from the kids. Because they only indulge a few times per year, the anticipation of something different is an aphrodisiac for both of them.

When it comes to butts, honesty is key. Because talking about butts tends to be considered taboo, many couples skirt the issue, which leads to misunderstandings, confusion, relationship turmoil, and barriers that get in the way of a fulfilling sex life. Not to be crass, but I suggest we spread open those butt cheeks and peer between them to see what comes up. Let's crack this issue wide open and tell the truth.

What do you think about anal sex? What are the risks? Why do so many men want it these days?

It really doesn't matter what *I* think about anal sex. Let's flip the question around and ask, "What do *you* think about anal sex?" Because what I think—or what anyone else aside from your partner thinks—is unimportant. With that caveat shared, here are a few thoughts about anal sex.

Many of us grew up believing that our anus is a nasty place. When we were babies, many of our parents said "yucky" every time they cleaned our bottoms, encouraging us to be potty trained as soon as possible. Once we were potty trained, Mommy was likely to say, "Oooh—stinky," as she flushed down the fruits of our potty labor. As a result, many people associate negative emotions with the anus, feeling a sense of conflict if they discover that anal stimulation excites them.

If you're a fan of anal sex or are tempted to try it, it's important to understand that it is not without significant risk. While anal intercourse has a lower risk of pregnancy, nearly all other risks are increased. Because the rectum hosts organisms that may exist nowhere else in the body, penetrating, licking, or manually stimulating the area in and around the anus can lead to infection.

If your mouth comes in contact with the anus, you can become infected with amoebas, parasites, and GI bacteria, such as shigella, *E. coli,* and salmonella. While these types of infections can usually be avoided with careful food preparation and handwashing, putting your mouth in contact with the anus essentially mainlines these potentially dangerous organisms into your GI tract.

Anal sex also leads to higher risk of sexually transmitted infection. Because tears in the sensitive rectal mucosa leave it susceptible, many STIs, including HIV, transmit more easily through anal intercourse. Diseases such as herpes and HPV can infect the anus and rectum, leading to anal warts and ulcerations. You can decrease the risk of these infections by using condoms, but some STIs, such as HPV and herpes, can still be transmitted in spite of consistent condom use. Make sure you use a silicone-based or water-based lubricant if you are using condoms. Oil-based lubricants, such as Vaseline or olive oil, can cause the condom to break down. Keep in mind that even if you're diligent about protecting yourself, condoms are more likely to tear during anal intercourse.

Anal sex can also cause physical damage to the rectum, increasing the risk of hemorrhoids, fissures, rectal prolapse, and fecal incontinence.[1] Infection of the anus with HPV can also increase the risk of anal cancer. I don't mean to scare you out of sexual bliss, but it's important to understand the risk you assume by engaging in anal sex.

Why do so many men want it these days? I honestly can't say. From talking to women, I get the idea that men tend to enjoy this experience more than women. But there are definitely women who enjoy it, too.

I've never had anal sex, but I want to try. Do you have any tips on how best to approach anal sex?

There's never a "right" or "wrong" way to relate sexually with your partner, as long as you're both on board. But yes, there are safer, more comfortable ways to engage in anal sex.

A FEW TIPS

1. Hop in the bath or shower first to make sure you feel clean and confident. While you can never completely remove fecal organisms from the rectum, you can reduce the number of bacteria right around the anus by using soap and water to clean yourself.

2. If you're worried about mess, put a towel underneath you. While you can use an enema beforehand to evacuate the contents of the rectum, putting liquid inside just before anal sex may make you more likely to leak.

3. Before you get started, visualize yourself having anal sex and enjoying it. If you can't dream it, you can't do it. You have to want it and you have to be able to envision it happening successfully. Your mind is more powerful than you think.

4. Use *lots* of personal lubricant. Water-based lubricants may dry up, but silicone-based lubricants tend to stick around. Advocates of anal sex say that if you're using too much lubricant, you're using almost enough. Unlike the vagina, which produces natural lubrication, the rectum does not self-lubricate. Penetrating the anus without lubrication can injure the rectum and anus—not to mention that it will likely hurt like hell.

5. Massage the anal sphincter to help the anus relax and allow entry, since contraction of the sphincter muscle can cause intense pain during penetration.

6. If you enjoy the feeling of anal stimulation, start small and gradually increase the size of what penetrates you. Experiment with a finger or a small sex toy designed for anal penetration.

7. Use condoms to cover the penis or sex toys. The risk of STIs is much higher with anal sex. And even if you're sure you're both safe from STIs, men can get infections from exposing the penis to fecal bacteria. Plus, it makes cleanup easier.

8. Once you're absolutely sure you're ready, make sure your partner enters you very slowly, allowing your anus to accommodate gently. Try to relax, rather than tighten, your anal sphincter.

9. Make sure you're in control of the depth and speed of penetration. Your partner may get excited, but feel free to put on the brakes. Remind your partner that you are not a porn star. If porn stars are the professional athletes of sex, you may be as rookie as they come. While the backdoor porn flicks may show guys humping with abandon, you're not going to want that when you're starting out.

10. Bear down around your partner when you're being penetrated. It helps to relax the anal sphincter.

11. Stimulate your clitoris or have your partner help out during anal penetration. Not only will it distract you from what's going on back there, it will help you relax and have fun also.

12. If you feel pain, stop. Remember, you're in charge.

13. Once the anus has been penetrated, whether it's with a finger, a penis, or a sex toy, make sure not to cross-contaminate the vagina with fecal organisms. Once something has been in the anus, you must wash it thoroughly before you stimulate other sensitive parts of the body. Otherwise, you may end up with a raging vaginal infection.

14. Make sure you avoid inserting any object into the rectum that does not have a flared end or stopper, which helps prevent it from being sucked into the rectum, where objects can be trapped. As someone

who has had to retrieve several objects from patients' GI tracts as a medical student, I can tell you it's not much fun for anyone.

My boyfriend wants me to strap on a dildo and have anal sex with him. Does that mean he's gay?

Not at all. While many link anal sex with gay men, many straight men also enjoy anal penetration. Penetration stimulates the prostate, which has been referred to as the "male G-spot." If you don't feel comfortable with the dildo and you want to help your partner experiment with anal penetration, you can use your finger or sex toys designed specifically for this purpose. A special toy, called a "butt plug," can be inserted into the anus and enjoyed during other sex acts, or you might try a rectal vibrator. But if the whole thing makes you feel uncomfortable, it's okay to say no.

My boyfriend wants to have anal sex and I'm curious but scared. What can I expect if I say yes?

I have to admit that I've never felt called to try anal sex. I guess for me, it's just an "Exit Only" kind of place, so I can't comment personally on what it feels like. To answer your question, I had to ask around. Here's what a few of my friends have to say:

> At first, it hurt, just like the first time I had vaginal sex. Now, it doesn't hurt anymore. It feels like lots of tingles and a deep sense of connectedness with my partner.

If you can surrender into the experience, really let go into it and relax, anal sex can be lots of fun. I don't orgasm from anal sex alone, but the anus is surrounded by very sensitive nerve endings and it definitely turns me on. If you do it right, it shouldn't hurt at all. The key is to use lots and lots of lube and to go very slowly in the beginning. Don't try to fight it. Relax all the muscles and enjoy.

My boyfriend really loves anal sex, and I consider myself sexually adventurous, but no matter what I do, anal sex hurts. I've read all the advice, tried all the tips, and it just doesn't work for me. Frankly, I hate it, and I'm not about to do something I hate. My boyfriend finally gave up.

Anal sex feels like being filled and stretched at the same time. The anus is tighter than the vagina, and having a vibrator in my vagina and a penis in my anus, or vice versa, both moving differently is highly erotic. I feel out of control. And because my partner is doing it, the power I have given away and the trust I put in him is part of what turns me on.

Ick. Yuck. Ouch. Poo explosion. Pain. Embarrassment. Humiliation. Need I say more?

It's not a regular thing for me and my partner, but I love the adventure of anal sex. I guess it's sexy, in part, because we only do it on special occasions. Because many consider it taboo, it feels a bit naughty, and since I'm kind of a good girl in other ways, that turns me on. It doesn't feel like stimulating the clitoris or the vagina, but in a funny sort of way, it feels like both at the same time to me. If you stimulate the clitoris while you're having anal sex, whoa, Nelly!

Nobody can ever help you predict what your individual experience will be like. If you're curious, you'll just have to see for yourself.

How common is anal sex?

The popularity and frequency of anal intercourse appears to be cultural. While Alfred Kinsey found that approximately 40 percent of people in the United States had engaged in anal intercourse,[2] only 3.5 percent of Koreans admitted to engaging in the act.[3] More recently, the Durex Global Sex Survey revealed that 55 percent of Greeks and 1 percent of Taiwanese engaged in anal sex. Some cultures routinely practice anal intercourse, either as birth control or as a way to protect a woman's "virginity," while some countries (and even some states in the United States) have sodomy laws, making it illegal to engage in anal intercourse, even between consenting adults. While anal sex is common, you're still in the majority if you've never had it. Some people dig it. Other's don't. There's no reason to judge it either way.

Anal sex sounds completely disgusting to me, but I don't want to be a prude. Am I missing something? Should I just suck it up and give it a try?

Absolutely not. There is nothing wrong with you if you don't want to try anal intercourse. Don't ever let anyone pressure you into trying something you don't want to do. Many couples never try anal sex, and there's absolutely nothing wrong with that. Like all sex acts, it must be something you really want to do, not something you feel you should endure to please your partner. It's important to be honest with your partner about how you feel. Remember, it's your body. You're in charge. If you don't want to have anal sex, skip it.

Does having anal sex give you a stretched-out butthole and make it so you can't hold in your farts and poop?

Not usually, but it is possible. Most of the time, the anus and rectum fare just fine. When the anal sphincter relaxes gently, even large things can be inserted into the rectum without trauma. Just look at the size of some of your poo. When you move your bowels, your anal sphincter opens, releasing the contents of the rectum gently and painlessly.

Most people can have anal sex without suffering any complications. Many of my patients have had regular anal sex for decades and seem just fine, although some tell me that having a poo feels different than it did before having anal sex, as if the poo just falls out of the butt. I have to admit that there does seem to be a difference between those with exit-only anuses and those with revolving doors. Things are just—how shall I say it?—a bit more spacious when I perform rectal exams.

If the anal sphincter is being forced open, damage can occur and increase the risk of leaking poo or being unable to control farts. The risk is probably greatest when aggressive anal sex tears the muscle around the anus. A famous porn star recently let one rip on a reality TV show, and when I got wind of that (no pun intended) I couldn't help wonder whether the poor thing can't help it after all that backdoor action.

If my partner wants to lick my butt, is that safe? How do I keep it clean?

Analingus, also called "rimming," refers to oral stimulation of the anus, usually by licking in and around the anus. Some men and women enjoy having the sensitive area around their anus stimulated, usually in combination with other types of oral sex. Even more so than with other types of oral sex, analingus requires good hygiene, since this area often carries harmful organisms not meant to be ingested. If you plan to engage in this behavior, clean around the anus with soap and water, understanding that, even with good hygiene, diseases may be transmitted, especially the types of gastrointestinal organisms that spread from fecal-oral routes, such as giardia, salmonella, shigella, and hepatitis A. You can decrease this risk by using a dental dam, a rectangular piece of latex used in dentistry.

I've heard poop sometimes comes out of the vagina. Can this really happen?

Yes. Unfortunately, this can happen if a woman develops a recto-vaginal fistula, a hole between the rectum and the vagina. Usually, recto-vaginal fistulas develop as the result of injury during childbirth. When I was working with refugees from Africa, fistulas that resulted as complications from obstructed childbirth were not uncommon. In developing countries, where access to C-section for obstructed labor is less available, these kinds of fistulas usually go hand in hand with stillbirth. When this occurs, holes between the bladder, the urethra, the vagina, and the rectum (or all of the above) may develop. Urine or fecal material can pass through the holes and wind up in the vagina, leading to chronic infections and other complications.

Sadly, the plight of women with fistulas in some developing countries is horribly sad. The World Health Organization estimates that 2 million women remain untreated. These women are often young teens whose families arranged for them to marry older men the women didn't even know. A childbirth complication like this may result in the woman being abandoned by her husband and ostracized from society because of her affliction. It breaks my heart.

What causes hemorrhoids? What are they?

Hemorrhoids are swollen, inflamed veins that protrude from around your anus and can be buried inside the lower rectum as well. By the time we reach the age of fifty, about half of us have experienced hemorrhoids. Just like you can get varicose veins in your legs, you can get varicosities in your butt. Hemorrhoids are particularly common during pregnancy, when the pregnant uterus obstructs venous return and allows blood to pool in veins in the lower part of the body.

In fact, if you examine the butts of women right after they push a baby out, many of them have some evidence of hemorrhoids, which usually resolve within six weeks postpartum. Other predisposing factors include chronic constipation or diarrhea, straining during bowel movements, prolonged sitting, obesity, and anal intercourse.

How come I sometimes find blood when I wipe my butt?

It's never "normal" to find blood when you wipe, but it is not uncommon. One study found that 13 percent of people occasionally find blood on the toilet paper when they wipe.[4] Causes include hemorrhoids, anal fissures, polyps, rectal ulcers, and, rarely, cancer. If you see blood in your stool, in the toilet after moving your bowels, or on the toilet paper after you wipe, chances are that it's nothing. But it's not something you want to ignore. Sometimes that's the only signal of a colon cancer lurking within, so please get it checked.

CHAPTER 15

You and Yoni:
The Relationship

I'M ASHAMED TO ADMIT THAT while I may be a gynecologist and you might consider me something of a vagina expert, my personal relationship with the body part that defines me as female is relatively new. Like many of you, I have always hated the word *vagina*. Because my father was a physician, we kids grew up using the proper names for all body parts. Heading off to the potty at the age of six, I was likely to utter something like, "I'm heading to the toilet to urinate and move my bowels. Mom, would you please help me wipe my anus?" Well, maybe not quite so bad, but you get the picture. There's was no talking about number one or two, and the vagina was definitely a vagina, not a pee pee or a front bottom.

Sex ed happened early, so I knew that babies were made when the daddy's penis enters the mommy's vagina and the daddy ejaculates the sperm, which then meets up with the egg. I laughed when other kids said that babies came from the mommy's stomach. I knew better.

Adolescence shifted the relationship, as my vagina became this mysterious, curious orifice I didn't understand and preferred to keep under wraps. Not until my teen years, when guys started expressing interest in poking around down there, did awareness of my own girl parts emerge. Discovering that this secret spot was capable of orgasmic pleasure changed everything, but with the ecstasy came shame, embarrassment, and guilt. Suddenly, the vagina was trouble.

Although my family talked openly about sex and the body, it was clear from the get-go that my vagina was expected to remain chaste until I was deflowered on my wedding night. Which isn't what happened. My deflowering happened on a twin bed in a smelly college dorm one day when my boyfriend's roommate was out for the night. While orgasms were nice, it turned out my vagina didn't like sex one bit. It hurt like the dickens, which defined my relationship with my vagina for the next decade.

When I got married in my twenties, things only got worse. Instead of squealing with pleasure, my vagina clamped down like a vice, closing out my husband and making sex nearly impossible. I became completely dissociated from what was going on down there. During sex, I would think about cooking, painting, lying on a lounge chair on the beach—anything to distract my mind from the pain. Over time, I lived more and more in my mind. My body barely existed, serving only to transport me from point A to point B.

It wasn't until my thirties, after the painful sex problem resolved, that I started hearing rumblings from my nether regions. This coincided with meeting a man who treated me like a goddess and worshiped at the temple of my body. It was around that time that I first discovered the term *yoni,* the Sanskrit word meaning "source of life," which is also used to describe the female genitals. The minute I heard the word *yoni*, something within me lit up. I began calling my girl parts Yoni, and doing so brought her to life. No longer was this body part scary, icky, pain invoking, or silenced. Instead, Yoni was sassy, confident, brash, and a little naughty. I visualized her as this little cartoon character, a plump, pink uterus figure with stick arms and legs, a big mouth, and a spunky Queen Latifah attitude.

Shortly after that, Yoni and I began chatting, sometimes about sex but mostly about babies. I tried to shut her up, but she refused to be silenced any longer. She had things to say, and she insisted on being heard. Our conversations opened up a whole new dialogue between me and my body, which continues to shift to this day. Getting pregnant and giving birth changed everything. I developed a profound awe for the magic that lies within me, the possibility inherent in Yoni. I started to view her with mystical, curious, fresh eyes. But part of me still resists her.

They say, "What we resist persists." Well, here I am, writing this book, while Yoni screams deafeningly, yearning to be heard, witnessed, and understood. To write this book, I listened to your questions, interviewed experts, and got curious. Sexperts and gurus seem determined to get Yoni and me talking on a deeper level. Somehow, I get the feeling a new relationship will evolve. Yoni, what do you think?

Your story may be vastly different from mine, which is to be expected. This is personal stuff we're talking about here. When it comes to our relationship with the sacred feminine within us, things get complex. I no longer think of my vagina as merely an organ. Now I see her more as a vessel, ripe and fertile, filled with potential that goes way beyond the ability to make a baby. I see her as an energetic place, a wise, knowing, loving place. I know that she is no longer separate from me—she is me, and we are one.

I can't even stand to say the word vagina. *Everything about it grosses me out. How do you make peace with your vagina and stop feeling uneasy at the mere mention of the word?*

I hear you, sister. Rachel Carlton Abrams, M.D., sexuality workshop leader and coauthor of *The Multi-Orgasmic Woman: Sexual Secrets Every Woman Should Know* and *The Multi-Orgasmic Couple: Sexual Secrets Every Couple Should Know,* offers this wisdom: "Vagina comes from the Latin word for 'sheath for a sword,' so it is not necessarily the most powerful metaphor for, arguably, the most powerful and creative part of the human body. My general feeling is, if you don't like the word, toss it. Make up another one."

Most of us are raised to think of our girl parts as taboo. Off-limits. Dirty places destined to get us in trouble. Even well-intentioned parents hoping to protect us from teen pregnancy and sexually transmitted infections coach us to fear the vagina as a black hole of desire that will be the death of us. It's no wonder we grow up feeling yucky about *down there.*

Realize that your vagina is just another part of *you.* It's not some disembodied, disgusting dark pit that's separate from you. It *is* you, just like your heart is you, your lungs are you, and your eyes are you. If you hate your vagina, you're hating you, and that's never healthy.

Explore this part of your body so it doesn't seem so foreign. Investigate the negative thought patterns that run like ticker tape in your mind. Try to turn these thoughts around by stating positive affirmations about your vagina, such as, "My yoni is the giver of life," or, "I am one with my girl parts." Merely setting the intention that you wish to make peace with your privates begins the process.

Dr. Abrams says, "It is very difficult to enjoy one's genitals if you think they're gross, and let me tell you, they are far from gross from a human perspective. The female genitalia have been celebrated by artists, priests, and lovers for as long as we have been on earth. Many times, we are uncomfortable because we are unfamiliar with our sexual organs or we have been taught to be ashamed of our sexual selves. Reclaiming the power and beauty of your—you name it—is immensely liberating. Women, and the sexual distinctions that make us women, are the most powerful creative forces in the world."

How come the prospect of going to the gyno makes me so nervous? I mean, gynecologists see vaginas all day long. No big deal, right? Then why is the experience so embarrassing, like I have the only vagina in the world?

Vaginas are not used to coming out in public. It's not that they're inherently bashful—in fact, they're arguably our most brazen body part. But we've closeted poor Yoni for so long that it's no wonder she's a little hesitant. And come on—baring yourself for the gynecologist isn't exactly the same as letting your full monty shine on some beach in the south of France. We're talking cold metal duckbill here. Stretching, scraping, and smashing. It's not exactly Yoni's idea of heaven.

I'm a gynecologist, and I still get embarrassed seeing the gynecologist. After all, these doctors are my colleagues. We see one another in the doctors' lounge between surgeries, and then there I am, butt naked, with my crotch in the face of someone I do C-sections with. Frankly, it's something I'd rather avoid.

Maybe it's not so bad that we protect the vagina. No, you're not the only woman in the world with a vagina, but vaginas are simply different than big toes are. We imbue Yoni with a sense of the sacred, and why not? She certainly deserves our respect. While the big toe brings balance to our gait, Yoni brings us pleasure and life. No comparison, right? While I don't think you should ever feel ashamed of your vagina, there's nothing wrong with guarding her closely.

That said, when it's time to go to the gynecologist, coax Yoni out gently. Remind her that we gynecologists specialize in vaginas. We see them every day. We don't judge them, and really, it's all very clinical to us. While Yoni may be more sacred than the big toe, to us gynecologists there's honestly not much difference.

So spread 'em, breathe, and visualize a pink bubble surrounding you, with a rosebud where your vagina is. When your gynecologist sits down on that stool, just let the rosebud open, petal by petal. Remind Yoni that you're nurturing her by seeing the gynecologist—that you're simply watering and pruning the rosebud. Then, as soon as the gynecologist snaps the speculum shut, close that rosebud and return Yoni to the sacred place where she feels more comfortable.

I feel uncomfortable with all things sexual and get all wigged out whenever it seems inevitable. What's my problem and how can I relax during sex?

Most of us have sexual hang-ups, some warranted, some not. If you've been raised to believe that you should banish the temptress within you and that sex is sinful, it's no wonder you get wigged out.

Maybe you've been the victim of rape or sexual molestation, which rocks your sexual foundation. Perhaps you barely know your sexual partner and don't feel comfortable making yourself so vulnerable, or perhaps you know your partner well but don't feel safe. Maybe you don't feel comfortable in your own skin. If any of this sounds familiar, talk to a therapist. You deserve to process those experiences now, or they may continue to haunt you.

While hesitation about sex can show up to protect us, sometimes it just gets in the way of sexual bliss. Here are a few tips to help you relax into your sexuality:

1. **KNOW AND LOVE YOUR BODY.** If you're not celebrating your body, it may rebel and put up red flags during sexual activity. Love yourself, just the way you are.

2. **DO A BODY BLESSING EVERY NIGHT BEFORE YOU GO TO SLEEP.** Close your eyes and scan through your body parts, starting at the top of your head and working downward. As you approach each body part, pay attention to any negative thoughts that come into your mind, such as, *I hate my fat ass.* Turn it around into an affirmation, such as, *Thank you, butt, for cushioning me when I sit.* Do this all the way down to your toes.

3. **TRY RELAXATION EXERCISES BEFORE YOU ENGAGE IN SEXUAL ACTIVITY.** Take a warm bath or have your partner massage you. Close your eyes and take a few very deep, cleansing breaths together. Light candles and incense; turn on soft, tinkly music—whatever it takes to help you relax.

4. **GO SLOW**. If you're feeling uncomfortable or nervous during sex, perhaps you're not ready. If your partner is truly caring, he or she will not pressure you. Give yourself permission to set the pace.

5. **MASTURBATE**. If other sexual activities make you nervous, try pleasuring yourself instead. The more you can relax into your own sexuality, the more you can comfortably share it with someone else. If you're able to masturbate by yourself without feeling uncomfortable, try masturbating together. That way, you have control of the whole experience, but doing it together can be highly stimulating for both of you. Once you graduate from that step, you may find yourself more comfortable with other sexual activities.

6. **BREATHE DEEPLY DURING SEXUAL ACTIVITY**. Deep breathing relaxes your muscles and alters your mind. Pay attention to your breathing throughout the whole experience.

7. **BE SILLY**. Sex doesn't have to be so serious. What better way to release the nervousness than to laugh? Laughing releases endorphins and stimulates the body in a whole host of ways. Try some laughter yoga exercises. Here's one: Stick your thumbs in your ears and stick your tongues out at each other. Now laugh out loud as hard as you can. Sounds goofy, I know, but try it! I swear it works.

8. **TALK ABOUT HOW YOU FEEL**. Don't be nervous in silence. If you're feeling nervous, talk about it. If you don't feel comfortable talking to your partner, talk to

a girlfriend. Don't judge yourself for your nervousness. In time, this, too, shall pass.

9. **ACKNOWLEDGE YOUR FEARS.** What is making you nervous? What lies at the root of your feelings? Are you living in the present moment, or is your past getting in the way? Write your answers in a journal.

10. **ENGAGE IN ACTIVITIES THAT MAKE YOU FEEL EMPOWERED.** Train for a race, donate your time to a charity, campaign for a cause you believe in, or do a health cleanse. For more ideas about how to feel empowered, check out OwningPink.com. Feeling empowered makes you feel more secure in vulnerable situations. Anything you can do to increase your sense of empowerment will follow you into the bedroom.

Why does my mood take a sudden 180 after having random sex with some hot guy? It goes from, Oh, what the hell, you only live once! *to,* Oh my God, what was I thinking! Now I probably have an STD!

I hear you. When you're dancing butt-cheek-to-butt-cheek in a sweaty club after a few drinks, you may be tempted to live on the edge. You might follow your lust back to his apartment, leaving your inhibitions at the door. Then, an hour later, after an unprotected romp in the sack, your margaritas begin to wear off and your logical brain springs into action, screaming, *What the hell have you done?*

One night, I was home sleeping, when my pager blared at 3 A.M. When I dialed the number, all I heard was sobs. Finally, Laura was able to pull it together and say, "I just had unprotected sex in Vegas with some guy named Vince."

If you find yourself feeling like Laura, you're not alone. Impulse sex may feel fun in the moment, but it rarely satisfies and may come back to haunt you in the form of an unwanted STD or pregnancy. And then there's the emotional junk that comes with it—the way you devalue yourself and your body when you aren't selective about who you choose to sleep with.

Sure, some women are Samantha from *Sex and the City*. They can hop in bed with hot guys, have meaningless sex, and shrug the whole thing off while they're zipping up their skirt. But most of us carry more baggage after a roll in the hay.

If you're regularly feeling regretful after sex, maybe it's time to reconsider how you decide who to sleep with. Once you're in a safe, loving relationship, chances are that you won't hear the ugly voices in your head. You'll just curl up in the arms of the person you love and feel grateful that your days of regret are behind you.

Some women seem like they just radiate sexy, but I'm not one of those women. How can I change that?

I've always felt the same way, like I'm more cute than sexy, but Regena Thomashauer, Sister Goddess, author, and founder of Mama Gena's School of Womanly Arts, thinks differently. She says:

> If you were born female, you *are* sexy. That's the deal—you can't do anything about it. You just are. You have all the equipment. It's your birthright. How to step into the experience

of that is a question of ownership. If you have a dusty old piano that nobody plays, there's no music. But you can start one key at a time, just like a piano lesson. Then, you can slowly expand your skill set so you can own the symphony that you are.

Every woman is beautiful, but if you don't believe it, you'll never fully step into that beauty. Owning your beauty is an inside job. It's a practice. There are little tiny things that make the difference between a woman who feels hot and a woman who doesn't. Even if I'm going to the gym, I'm going to put on lip gloss because it will change my experience of myself between home and the gym. I know that—so why would I deny myself the confidence I derive from taking four seconds to make myself feel sexy? Every woman has her own journey and is already equipped with tips and secret steps she will take to feel beautiful. Once you know what's in your toolbox, it's there for you to use anytime.

How come I often feel gross and ashamed right after masturbating? What can I do to fix it?

Many of us are raised to believe that masturbation is shameful. When we're young and exploring, our parents might yell, "Judy, get your hands out of your pants! That's nasty." Our teachers might send us to time-out. And it may get worse as we get older. It's appropriate that we learn that masturbation is best enjoyed privately. But along with learning to hide it, we pick up vibes that lead us to believe that it's wrong when there's absolutely no reason to feel ashamed, gross, or naughty after masturbating. It offers a healthy, safe release of sexual tension and may even improve your health and your sexual life.

If you feel icky after masturbating, try this yoni meditation.

An Exercise for Learning to Love Your Yoni

Caroline Muir, founder of the Divine Feminine Institute and coauthor of *Tantra: The Art of Conscious Loving,* recommends using this exercise to strengthen the relationship between you and your yoni:

Make a meditation practice of placing your right hand over your yoni and your left hand over your heart. Focusing on your breath, fill up your heart center with a long, deep, complete breath. Exhaling, breathe love out through the hand resting on your yoni . . . breathe your yoni's energy up to your heart, filling your heart with your aliveness and life force. Keep this circuit of breath and energy going for several minutes before letting go of technique and feeling into the connection between your passion and your love.

Next, look at your yoni in a mirror. Examine yourself in a non-aroused state, exploring with the curiosity of a child. Looking into your own eyes, see the perfection of how you are created. Remember that the Divine doesn't make mistakes. You are perfect, just as you are. Be patient with yourself. No one taught you how to love yourself, and especially how to love your yoni.

Then begin to arouse yourself, with love. Marvel at the color of your labia, the shape of them, the changes in your clitoris as you arouse this tiny organ. Breathe your awakening pleasure up through your heart, breasts, and into your brain. Exhale with sound, sounds of joy or pleasure or any sound meant to awaken you. If you feel frustrated, make a frustrated sound. Joyful? Make a joyful one. Angry? Express it.

Then, when you feel ready, lubricate a finger and slowly enter your own yoni in a way that feels honoring, respectful and loving. Curl the finger up and stroke the upper, forward wall of your yoni—where your sacred spot, your G spot, lies. It may take time for you to feel your pleasure. You may be numb or you may feel burning inside. Be patient. Love slowly, gently, as if you were coaxing a rosebud to open to its beauty and fragrance. Do the same with your clitoris. It should not be forced to "perform"—just enjoy the journey of loving yourself. There is no goal other than to affirm love and acceptance of your sacred temple of love.

Eventually, orgasms will happen easily. Women are capable of limitless orgasmic pleasure, free flowing and abundant. That is our birthright. Breathing and relaxing all of the genital and leg muscles will allow pleasure to move freely. Tension and breath holding are what block full pleasure from moving freely. When you release in love, you pave the way for heightened sexual healing, pleasure, intimacy, and communion.

I have to get a hysterectomy. Will it make me less of a woman?

Absolutely not, but you're not alone for feeling like it might. It's funny when you think about it. The uterus. It's just a small muscle, shaped like a piece of fennel, lined with glands and topped off with two squiggly fallopian tubes like Medusa hairs. At the root of the uterus lies the cervix, the mouth of the uterus, which dilates to let a baby out. The uterus is certainly not the heart or the brain. Technically, it's more like the gallbladder. While it has its function, you can certainly live without it.

But somehow, the uterus and the gallbladder are different. Many people happily relinquish their ailing gallbladders, but for many women the uterus represents a physical manifestation of the sacred feminine—a spiritual vessel, ripe with possibility. When you are young, the uterus represents all the lives you might bring into the world, with all its nourishing tissue ready to embrace and sustain another being. It signifies all the sexual energy you exude into the world. But even beyond that time, when the ovaries have run out of steam and pregnancy is unlikely, women cling to the uterus with bittersweet attachment.

I remember the day I told my patient Jeanine she needed a hysterectomy. Jeanine suffered from severe endometriosis and kept showing up in emergency rooms, hemorrhaging and requiring blood transfusions. She suffered from pelvic pain for many years, and I had performed several endoscopic surgeries to look at her pelvis, treat cysts on her ovaries, and remove polyps from inside her uterus. We tried every conservative therapy I could think of, plus she saw an acupuncturist and a Reiki practitioner. Nothing worked. Jeanine was literally bleeding to death.

Now, I'm not one of those doctors who rush to remove every uterus. When I was training to become a doctor, we joked about one doctor who took out a lot of perfectly normal uteruses for no apparent reason. Our tongue-in-cheek diagnosis was "chronic persistent uterus." I have never been that way. Most of my patients are begging for a hysterectomy by the time I finally agree to perform one.

But Jeanine was not begging. She was crying. She touched her abdomen, as a pregnant woman would caress her round belly. "But I'll miss her," she said.

I've discovered that women have complex relationships with their uteri. Can't live with 'em, can't live without 'em. While we might gripe, you'd better not go dissing our girl parts. It's akin to having some stranger insult your crazy uncle Denny. If he's your uncle Denny, you can prattle on about his pet snake and the weird ferret who sleeps with him at night, but if anyone else tries to attack Uncle Denny, you'll bare your fangs. That's how many women feel about their reproductive organs. They bitch and moan about how it sucks to be a woman, but if you tell them it's time to say good-bye to Aunt Flow, they cling with rapt affection.

Before I took out Jeanine's uterus, I reminded her to remember that the true essence of her womanhood lies in her spirit and can't ever be taken away from her. While she may grieve the loss of her uterus, she will still be all woman, through and through.

Caroline Muir says, "Hysterectomy does alter the body and certainly may affect orgasmic response; however hormonal issues and emotional self-deception are the only things that can 'make you feel like less of a woman.' You are capable of increasing pleasure throughout the course of your life when you bring love and healing to these misunderstood or surgically impacted aspects of your sexuality."

Is it possible to cause gynecologic problems just because I hate being a girl and despise my vagina?

I'm a big believer in the mind/body/spirit connection, so if you go around rejecting your girl parts, they just might reject you. I'm not suggesting it's a tit-for-tat thing. It's not like if you curse your uterus, she's going to instantaneously zap you with a menstrual cramp. But I do believe that our attitudes and beliefs can manifest as health issues.

Christa certainly believes that her gynecological problems have something to do with how much she has always resented being a girl. Early on, Christa decided that having female reproductive organs was not a blessing but a curse. Getting her period for the first time was one of the worst moments of her life. She was thirteen years old and babysitting at her neighbor's house when she started to bleed. All the other girls in her school couldn't wait to get a front-row seat on the blood train, but she cursed the day her womanhood started.

By the time she was in her twenties, her girl parts were starting to fight back. She told me this story:

"I remember making a painful trip to the ER in the passenger seat of my date's car. Talk about embarrassing. 'Um, yeah, great date, a lot of fun and all. But I think there is an alien in my gut, so, um, could you please take me to the hospital? *Now?*'

"I don't remember much from that day except the examination I received from a male gyno, who I'll call MG. Until this fateful day in the ER, all my gynos were women. As Murphy's Law would have it, I got a male doctor who obviously failed Bedside Manner 101 in med school. So here I am, grimacing in pain with my legs in stirrups, hoo-ha completely exposed and lit from above with what seemed like stadium lights from a football field. The whole bloody gynecological experience is so mortifying. I would rather have a root canal than be exposed, spread-eagle, to a complete stranger. In an effort to lower my blood pressure and inject some levity into this uncomfortable and highly embarrassing situation, I smiled at MG and said, 'Hey, if you're going to be down there, shouldn't we head out for drinks first?'

"With a stone-cold, judgmental face and reprimanding, condescending tone, MG barked, 'This is no time for jokes, young lady!'

"There I was, in the stirrups, thinking, 'That humor-deficient, stethoscope-wearing penis didn't just say that. Did he?'

"Apparently MG never read the *Female Jokester Chronicles* in medical school. If he had taken the time to check out chapter three, 'Humor Heals,' he would have understood a fundamental rule all women know—one of the best times to utilize humor is when your gynecologist has one hand shoved up your crotch and the other hand pressing down on a soon-to-be-bursting cyst. Duh!

"Over the course of several years and a great deal of pain, I had four surgeries for endometriosis. I spent hours of my life pissed off because I felt cursed. I was probed, prodded, poked, and processed in several Philadelphia-area hospitals. I asked God, 'Why me?' over and over and again.

"Recently my female problems started up again. I was in the parking lot of my gynecologist's office last week, and I had a huge ovarian mind-body-spirit revelation. This revelation didn't come easy. My 'aha' moment came only after I had a hissy fit and screamed a string of expletives at the sky. My rant would have made a longshoreman blush.

"Here's what I realized: All the pain, drama, and bullshit I experienced while walking around in this girly skin was caused by my very own thoughts. It is all so simple. I thought I was cursed—so my body cursed me. I thought the pain controlled me when, in fact, I created the pain. Aha!

"Thoughts create reality.

"Wow. Lightbulb. So now I resolve to:

"One: Accept the skin I am in.

"Two: Act in loving and nurturing ways toward myself.

"Three: Embrace my good health.

"Four: Raise a toast to my two ovaries.

"Five: Love myself unconditionally.

"Six: Nourish my female fabulousness."

Christa still hates all things gyno, but changing her state of mind has resulted in a shift. She hasn't had pain or surgery in years now.

If you have gynecologic problems and have cursed your womanhood, don't blame yourself. I'm not suggesting that it's all your fault. Many other factors come into play: genetics, environmental factors, hormones, and sheer bad luck. But if you focus on what's wrong with your body, rather than what's right about it, that certainly isn't going to encourage your body to heal. Take a moment to feel gratitude for all the beauty that comes with being female. After all, we girls get to wear sparkly, frilly pink things, carry new life inside us, resonate on a deeply emotional level, hold the family together, enjoy long lunches and phone calls with our girlfriends, and be cheerleaders for the world. All that, and we get to have boobs! So count your blessings. Being female is truly a privilege.

As I get older, I feel like the fresh young mining village that was my vagina years ago has become an old closed-down mining town. Is that all in my head? How can I change it?

Maybe it's time to do a little drilling and rock the socks off that mining town! After you blow the dust off yourself, take a moment to shift how you think about things down there. How you feel about your girly bits can certainly make a difference. Sure, your pubic hair may be peppered with gray and things may look a bit different as we age, but youth is a state of mind, even in the vagina.

Have you been programmed to believe that only young vaginas are valuable? Have you considered revitalizing the way you think about the mining village that is your vagina? Maybe it's time for renewal.

Regena Thomashauer, says:

A woman who owns her pussy owns her life. If you don't feel good about your pussy, you don't feel good about your life. If you feel fantastic about your pussy, you feel fantastic about your life. It's an opportunity not only to own your beauty, but to learn the journey and the experience of each of the pussy's eight thousand nerve endings and how that informs your being. If eight thousand desires, decisions, and dreams are not about pleasure, then you're not really living what it means to be a woman. The key is to be guided by the physiology and to pay attention to the song that your body wants to sing with you. Learn the poetry that your body wants to whisper in your ear. Open yourself to pleasure and rapture. It will transform you.

The genital tissue is elastic and luscious and responsive for your entire life. You can continue to expand sensually for your whole life. It's very good to put the key in your own ignition and then you can invite passengers. Women are obligated to do an enormous amount of discovery and self-exploration. Once you know how to dance, you can have a good ride on the dance floor.

If changing your thought patterns doesn't help, you could be responding to hormonal changes in your body. If you're feeling like your vagina is trying to close up shop, you may be noticing the effects of estrogen deficiency. To keep your vagina sparky and fresh, talk to your gynecologist about whether estrogen might help bring new life to the old mining town.

Check in with your body and celebrate the woman that you are. If you feel vital, saucy, and sexy, you'll have that mining town hopping before you know it.

I wish I felt like Samantha when it comes to my sex, but really I'm more of a Charlotte. How can I improve my sexual confidence?

I can relate. Honestly, I'm more of a Charlotte myself. I used to wish I were different—spicier, more daring, less cute, more va va voom. I've since realized that I'm never going to be a Samantha. Why try to be someone I'm not? It wouldn't be authentic to who I really am. But that doesn't mean I can't be the sexiest Charlotte possible.

Because of her boundless experience helping women embrace their sexual confidence, I asked Sheila Kelley, founder of S Factor, to chime in. Sheila says, "Don't try to be someone you're not. We are all idiosyncratically different, yet we all feel like we're supposed to be exaggerated, buffoony images of sexuality because we think that's what guys want. I'm here to tell you it's just not true. I spent a year going to strip clubs and watching the guys. What I discovered is that it didn't matter what a woman looked like, what image she portrayed, or how old she was. Guys at strip clubs were listening to the voice of how those women's bodies moved."

To help you embrace your sexual confidence, Sheila says, "find your feminine movement. Feel your body. Touch yourself. Move your hips in round circles in the privacy of your own home. As a society, we have turned off so many attributes of being feminine. Snap on those off switches, and you will discover your own confidence. Every woman has to figure out for herself what feels right. You can have intellectual confidence, physical confidence, emotional confidence, but if you don't practice the feminine movement muscle, you'll never optimize your sexual confidence. If you don't find it within your body, you'll always be *thinking* about sexual confidence, not *feeling* sexual confidence."

Even though my husband thinks I'm super-sexy,
I always feel bad about my body and my ability
to pleasure him. What can I do about that?

Why are we are own worst critics? We blame men for all the pressure we feel to be thin, beautiful, and sexy, but we women are just as hard on ourselves. I know I'm just as guilty. I get dressed up in lacy lingerie, hearing Marvin Gaye's "Let's Get It On" in my head. I start to wiggle a bit, feeling sassy and sexy. Then I catch a glimpse of myself in the mirror and I become a deflated balloon, with all the air squeaking out of me in one lame hiss. Or I'll be in the middle of going down on my honey when all of a sudden I get self-conscious and insecure. What if I'm not doing it right? What if he's comparing me to his ex-girlfriend? What if he's secretly fantasizing about Jessica Alba? What if he's just wishing I'd get it over with already so he could go to sleep? These intrusive thoughts are enough to knock the sails out of even the sexiest woman. Of course, if I talk about my insecurities with my husband, he looks at me like I have two heads. He thinks I'm gorgeous and sexy and have as many moves as I need to please him *and* his disco stick.

How can we expect to be sexy when we're busy obsessing about the cellulite on our thighs, our pathetically small (or monstrously ginormous) boobs, our sagging buttts, or the stretch marks on our bellies? Even those of us with seemingly perfect bodies will find minor imperfections to obsess over. My friend Laura Fenamore, founder of OnePinkey.com, used to be very overweight, with the crappy body image that tends to accompany obesity. Then, she lost over a hundred pounds. But after losing the weight, she still hated her body and realized that the issue was much more than skin-deep. A while later, she met a gorgeous woman with a perfect body, who, as it turns out, despised her body, too. The two women made a "pinkie promise" that they would start to love their bodies by loving just their pinkie finger. After all, what's not to love about a pinkie finger?

Maybe we can all start, one pinkie finger at a time. After you've made peace with your pinkie, your thumb, and your left elbow, try moving on to more difficult body parts, like your thighs, belly, and butt. How can we expect mind-blowing sex if we can't learn to love our own bodies? What kinds of messages are we sending to others in our lives if we constantly criticize ourselves?

Sisterhood is everywhere. Whether you find it in a knitting circle, a book club, a yoga class, a quilting collaboration, a women's retreat, a prayer group, an art class, around the dinner table, online, or at the gynecologist's office, you are not alone. If you don't have a support network, build one. Set the intention that you will attract like-minded people committed to giving and receiving love. Open your heart, and watch others flow toward you, as they hear the call that unites us. I am watching it happen around the globe, as the battle cry of sisterhood drowns out the loneliness, despair, violence, disconnection, and disease that plagues our society. I am witnessing transformation, not just in individuals giving birth to themselves but in a shift in the very vibration of the world. It warms my heart.

I know you may feel hesitant. When you're giving birth to you, you might feel as if you're jumping off a cliff to land on a balance beam that dumps you on a tightrope over a pit of hot lava. But I promise—it's not so scary. As you stand on that precipice, perched and trembling, you're about to find out that you can fly.

You, too, can find your mojo and give birth to yourself. Trust the process. Everything you need is already here.

"feeling my fur" in a whole new way, as I shimmy through life and learn to live lusciously in my skin. Mary Roach, thank you for making me laugh out loud with every communication. I'll just never think of sex research the same way after *Bonk*. Dr. Susanna Beshai, thank you for your friendship, research guidance, and trust.

And to all of you reading this book, thank you for trusting me with your most intimate parts. I hope you walk away feeling empowered down there. Within each of you lies a beautiful, radiant spirit full of mojo, and the first step to *owning* that is to love and honor your body.

Notes

2. HOW COOCHIES LOOK

1. R. L. Dickinson, "Hypertrophies of the Labia Minora and Their Significance," *American Gynecology*, September 1902.
2. Barry S. Verkauf, M.D., James von Thron, M.D., and William F. O'Brien, M.D., *Obstetrics & Gynecology* 80, no. 1 (1992):41–44.
3. Robert Latou Dickinson, *Atlas of Human Sex Anatomy*.
4. L. Miller and M. Edenholm, "Genital Piercing to Enhance Sexual Satisfaction," *Obstetrics & Gynecology* 93, no. 8 (1999):37.
5. A. Jemal, R. Siegel, E. Ward, et al., "Cancer Statistics, 2008," *CA: A Cancer Journal for Clinicians* 58 (2008):71.

3. HOW COOCHIES SMELL AND TASTE

1. S. Jacob, M. K. McClintock, B. Zelano, and C. Ober (2002). "Paternally Inherited HLA Alleles Are Associated with Women's Choice of Male Odor," *Nature Genetics* 30 (2002):175–179; C. Wedekind, T. Seebeck, F. Bettens, and A. J. Paepke, "MHC-Dependent Mate Preferences in Humans," *Proceedings of the Royal Society of London B* 260 (1995):245–249.

4. SEX AND MASTURBATION

1. Cheryl D. Fryar, M.S.P.H., Rosemarie Hirsch, M.D., M.P.H., Kathryn S. Porter, M.D., M.S., Benny Kottiri, Ph.D., Debra J. Brody, M.P.H., and Tatiana Louis, M.S., *Division of Health and Nutrition Examination Surveys Advance Data from Vital & Health Statistics, Drug Use and Sexual Behaviors Reported by Adults: United States, 1999–2002*, no. 384, June 2007.
2. http://www.womansday.com/Articles/Family-Lifestyle/Relationships/The-State-of-Our-Unions.html.
3. http://www.kinseyinstitute.org/resources/FAQ.html.
4. Barry R. Komisaruk, Beverly Whipple, Sara Nasserzadeh, and Carlos Beyer-Flores, *The Orgasm Answer Guide* (Baltimore, MD: Johns Hopkins University Press, 2009), 10–11.
5. Fritz Klein, M.D., *The Bisexual Option*, 2d ed. (Binghamton, NY: Haworth Press, 1993).

6. George Davey Smith, Stephen Frankel, and John Yarnell, "Sex and Death: Are They Related? Findings from the Caerphilly Cohort Study," *British Medical Journal* 315 (1997):1641–1644; E. Palmore, "Predictors of the Longevity Difference: A Twenty-five-Year Follow-Up," *The Gerontologist* 22 (1982):513–518; G. Persson, "Five-Year Mortality in a 70-Year-Old Urban Population in Relation to Psychiatric Diagnosis, Personality, Sexuality and Early Parental Death," *Acta Psychiatrica Scandinavica* 64 (1981):244–253.

7. S. Ebrahim, M. May, S. Ben, P. McCarron, S. Frankel, J. Yarnell, and S. Dave, "Sexual Intercourse and Risk of Ischaemic Stroke and Coronary Heart Disease: The Caerphilly Study," *Journal of Epidemiology Community Health,* 56 (2002):99–102.

8. M. G. Lê et al., "Characteristics of Reproductive Life and Risk of Breast Cancer in a Case-Control Study of Young Nulliparous Women," *Journal of Clinical Epidemiology* 42, no. 12 (1989):1227–1233.

9. Carl J. Charnetski and Francis X. Brennan, *Feeling Good Is Good for You: How Pleasure Can Boost Your Immune System and Lengthen Your Life* (Emmaus, PA: Rodale Press, 2001).

10. Carol Rinkleib Ellison, *Women's Sexualities* (Oakland, CA: New Harbinger Publications, 2000).

11. David Weeks and Jamie James, *Secrets of the Superyoung* (New York: Berkley Books, 1998).

12. Ellison, *Women's Sexualities.*

13. Erika L. Meaddough, David L. Olive, Peggy Gallup, Michael Perlin, and Harvey J. Kliman, "Sexual Activity, Orgasm and Tampon Use Are Associated with a Decreased Risk for Endometriosis," *Gynecologic and Obstetric Investigation* 53 (2002):163–169.

14. Winnifred B. Cutler, *Love Cycles: The Science of Intimacy* (New York: Villard Books, 1991).

15. Cutler, *Love Cycles*; Mary H. Burleson, W. Larry Gregory, and Wenda R. Trevathan, "Heterosexual Activity and Cycle Length Variability: Effect of Gynecological Maturity," *Physiology & Behavior* 50 (1991):863–866.

16. Ellison, *Women's Sexualities.*

17. A E. Sayle, D. A. Sevitz, J. M. Thorp, I. Hertz-Picciotto, and A. J. Wilcox, "Sexual Activity During Late Pregnancy and Risk of Preterm Delivery," *Obstetrics & Gynecology* 97, no. 2 (2001):283–289.

18. Helen Singer Kaplan, "Desire? Why and How It Changes," *Redbook,* October 1984, 58, as cited in Barry R. Komisaruk and Beverly Whipple, 1995; D. Shapiro, "Effect of Chronic Low Back Pain on Sexuality," *Medical Aspects of Human Sexuality* 17 (1983):241–245, as cited in Komisaruk and Whipple; Beverly Whipple and Barry R. Komisaruk, "Elevation of Pain Threshold by Vaginal Stimulation in Women," *Pain* 21 (1985):357–367.

19. Randolph W. Evans and James R. Couch, "Orgasm and Migraine," *Headache* 41 (2001):512–514.

20. David J. Weeks, "Sex for the Mature Adult: Health, Self-Esteem and Countering Ageist Stereotypes," *Sexual and Relationship Therapy* 17, no. 3

(2002):231–240; Pamela Warner and John Bancroft, "Mood, Sexuality, Oral Contraceptives and the Menstrual Cycle," *Journal of Psychosomatic Research* 32, nos. 4/5 (1988):417–427; Edward O. Laumann, John H. Gagnon, Robert T. Michael, and Stuart Michaels, *The Social Organization of Sexuality: Sexual Practice in the United States* (Chicago: University of Chicago Press, 1994).

21. Joseph A. Catania and Charles B. White, "Sexuality in an Aged Sample: Cognitive Determinants of Masturbation," *Archives of Sexual Behavior* 11, no. 3 (1982):237–245.

22. Charnetski and Brennan, *Feeling Good Is Good for You;* Weeks, "Sex for the Mature Adult," 231–240.

23. David Farley Hurlbert and Karen Elizabeth Whittaker, "The Role of Masturbation in Marital and Sexual Satisfaction: A Comparative Study of Female Masturbators and Nonmasturbators," *Journal of Sex Education & Therapy* 17, no. 4 (1991):272–282.

24. Weeks, "Sex for the Mature Adult," 231–240.

25. Peter Gardella, *Innocent Ecstasy: How Christianity Gave America an Ethic of Sexual Pleasure* (New York: Oxford University Press, 1985); Barbara Keesling, *Rx Sex: Making Love Is the Best Medicine* (Alameda, CA: Hunter House, 2000); Gina Ogden, "Spiritual Passion and Compassion in Late-Life Sexual Relationships," *Electronic Journal of Human Sexuality* 4 (August 14, 2001), http://www.ejhs.org/volume4/Ogden.htm (accessed November 2009).

5. ORGASM

1. J. L. Shifren, "Sexual Problems and Distress in United States Women: Prevalence and Correlates," *Obstetrics & Gynecology* 112 (2008):970.

2. E. O. Laumann, A. Nicolosi, D. B. Glasser, A. Paik, C. Gingell, E. Moreira, and T. Wang, "Sexual Problems Among Women and Men Aged 40–80 Years: Prevalence and Correlates Identified in the Global Study of Sexual Attitudes and Behaviors," *International Journal of Impotence Research* 17 (2005):39.

3. Shere Hite, *The Hite Report: A National Study of Female Sexuality* (Seven Stories Press, 2003); Robert W. Birch, Ph.D., *Pathways to Pleasure* (2000).

4. A. K. Ladas, B. Whipple, and J. D. Perry, *The G Spot and Other Recent Discoveries About Human Sexuality* (New York: Holt, Rinehart and Winston, 1982).

5. Elizabeth Stewart, *The V Book: A Doctor's Guide to Complete Vulvovaginal Health* (New York: Bantam Books, 2002).

6. Mary Roach, *Bonk: The Curious Coupling of Science and Sex* (New York: W. W. Norton, 2008), 198–199.

7. B. Whipple, B. Myers, and B. R. Komisaruk, "Male Multiple Ejaculatory Orgasms: A Case Study," *Journal of Sex Education and Therapy* 23, no. 2 (1998):157–162.

8.

9. Komisaruk, Whipple, Nasserzadeh, and Beyer-Flores, *The Orgasm Answer Guide*, 110.

10. J. C. Rhodes, K. H. Kjerulff, P. W. Langenberg, and G. M. Guzinski, "Hysterectomy and Sexual Functioning," *JAMA* 282 (1999):1934.

11. B. Whipple, G. Ogden, and B. R. Komisaruk, "Physiological Correlates of Imagery Induced Orgasm in Women," *Archives of Sexual Behavior* 21, no. 2 (1992):121–133.

6. DISCHARGE AND ITCHING

1. M. J. Godley, "Quantitation of Vaginal Discharge in Healthy Volunteers," *British Journal of Obstetrics and Gynaecology* 92 (1985):739.

7. PERIODS

1. A. M. Kaunitz, "Choosing to Menstruate—or Not," *Contemporary OB/GYN* 46 (2001):73.

2. D. N. Ruble, "Premenstrual Symptoms: A Reinterpretation," *Science* 197 (1977):291.

3. Christiane Northrup, M.D., *Women's Bodies, Women's Wisdom*, rev. ed. (New York: Bantam Books, 1998).

8. FERTILITY

1. M. J. Zinaman, E. D. Clegg, C. C. Brown, J. O'Connor, and S. G. Selevan, "Estimates of Human Fertility and Pregnancy Loss," *Fertility and Sterility* 65 (1996):503.

2. C. Gnoth, D. Godehardt, E. Godehardt, et al., "Time to Pregnancy: Results of the German Prospective Study and Impact on the Management of Infertility," *Human Reproduction* 18 (2003):1959.

3. C. Gnoth, E. Godehardt, P. Frank-Herrmann, et al., "Definition and Prevalence of Subfertility and Infertility," *Human Reproduction* 20 (2005):1144.

4. Devendra Singh et al., "Frequency and Timing of Coital Orgasm in Women Desirous of Becoming Pregnant," *Archives of Sexual Behavior* 27, no. 1 (1998):15–29.

5. Roy J. Levin, "The Physiology of Sexual Arousal in the Human Female: A Recreational and Progreational Synthesis," *Archives of Sexual Behavior* 31, no. 5 (2002):405–411.

6. V. Beral, D. Bull, R. Doll, R. Peto, and G. Reeves, "Breast Cancer and Abortion: Collaborative Reanalysis of Data from 53 Epidemiological Studies, Including 83,000 Women with Breast Cancer from 16 Countries," *Lancet* 363 (2004):1007; M. Melbye, J. Wohlfahrt, J. H. Olsen, M. Frisch, T. Westergaard, K. Helweg-Larsen, and P. K. Andersen, "Induced Abortion and the Risk of Breast Cancer," *New England Journal of Medicine* 336 (1997):81; G. Erlandsson, S. M. Montgomery, S. Cnattingius, and A. Ekbom, "Abortions

and Breast Cancer: Record-Based Case-Control Study," *International Journal of Cancer* 103 (2003):676; X. Paoletti and F. Clavel-Chapelon, "Induced and Spontaneous Abortion and Breast Cancer Risk: Results from the E3N Cohort Study," *International Journal of Cancer*, 106 (2003):270; G. K. Reeves, S. W. Kan, T. Key, A. Tjønneland, A. Olsen, K. Overvad, P. H. Peters, F. Clavel-Chapelon, X. Paoletti, F. Berrino, V. Krogh, P. Palli, P. Tumino, S. Panich, P. Vineis, C. A. Gonzalez, E. Andanaz, C. Martinez, P. Amiano, J. R. Quiros, M. R. Tormo, K. T. Khaw, A. Trichopolou, T. Psaltopoulou, V. Kalapothaki, G. Nagel, J. Chang-Claude, H. Boeing, P. H. Lahmann, E. Wirfält, R. Kaaks, and E. Riboli, "Breast Cancer Risk in Relation to Abortion: Results from the EPIC Study," *International Journal of Cancer* 119 (2006):1741; K. B. Michels, F. Xue, G. A. Colditz, and W. C. Willett, "Induced and Spontaneous Abortion and Incidence of Breast Cancer Among Young Women: A Prospective Cohort Study," *Archives of Internal Medicine* 167 (2007):814.

7. C. Tietze "Reproductive Span and Rate of Reproduction Among Hutterite Women," *Fertility and Sterility* 8 (1957):89; N. Laufer, A. Simon, A. Samueloff, et al., "Successful Spontaneous Pregnancies in Women Older than 45 Years," *Fertility and Sterility* 81 (2004):1328.

8. J. Grifo and N. Noyes, "Delivery Rate Using Cryopreserved Oocytes Is Comparable to Conventional in Vitro Fertilization Using Fresh Oocytes: Potential Fertility Preservation for Female Cancer Patients," *Fertility and Sterility*, 2009.

9. "Optimizing Natural Fertility," *Fertility and Sterility* 90 (2008):S1.

10. Roy J. Levin, "The Physiology of Sexual Arousal in the Human Female: A Recreational and Progreational Synthesis," *Archives of Sexual Behavior* 31, no. 5 (2002):405–411.

11. A. J. Wilcox, C. R. Weinberg, J. F. O'Connor, D. D. Baird, J. P. Schatlerer, R. E. Canfield, E. G. Armstrong, and B. C. Nisula, "Incidence of Early Loss of Pregnancy," *New England Journal of Medicine* 319 (1988):189.

12. R. Rai and L. Regan, "Recurrent Miscarriage," *Lancet* 368 (2006):601.

13. A. J. Wilcox, D. D. Baird, D. Dunson, R. McChesney, and C. R. Weinberg, "Natural Limits of Pregnancy Testing in Relation to the Expected Menstrual Period," *JAMA* 286 (2001):1759.

9. PREGNANCY

1. http://www.otispregnancy.org/pdf/hair_treatments.pdf.

2. J. Sasaki, A. Yamaguchi, Y. Nabeshima, S. Shigemitsu, N. Mesaki, and T. Kubo, "Exercise at High Temperature Causes Maternal Hyperthermia and Fetal Anomalies in Rats," *Teratology* 51 (1995):233.

3. http://www.otispregnancy.org/pdf/hyperthermia.

4. J. Kavanagh, A. J. Kelly, and J. Thomas, "Sexual Intercourse for Cervical Ripening and Induction of Labour (Cochrane Review)," *Cochrane Database of Systematic Reviews* 2 (2001):CD003093.

5. J. Kavanagh, A. J. Kelly, and J. Thomas, "Breast Stimulation for Cervical Ripening and Induction of Labour (Cochrane Review), *Cochrane Database of Systematic Reviews* 4 (2001):CD003392.

6. A. J. Wilcox, C. R. Weinberg, J. F. O'Connor, D. D. Baird, J. P. Schatlerer, R. E. Canfield, E. G. Armstrong, and B. C. Nisula, "Incidence of Early Loss of Pregnancy," *New England Journal of Medicine* 319 (1988):189.

7. S. Cnattingius, L. B. Signorello, G. Anneren, B. Clausson, A. Ekbom, E. Ljungen, W. Blot, J. K. McLaughlin, G. Petersson, A. Rane, and F. Granath, "Caffeine Intake and the Risk of First-Trimester Spontaneous Abortion," *New England Journal of Medicine* 343 (2000):1839.

8. http://www.otispregnancy.org/pdf/caffeine.pdf.

9. Elizabeth Armstrong, *Conceiving Risk, Bearing Responsibility: Fetal Alcohol Syndrome and the Diagnosis of Moral Disorder,* 1st ed. (Baltimore, MD: Johns Hopkins University Press, 2003), 72–93.

10. Y. Kelly, A. Sacker, R. Gray, J. Kelly, D. Wolke, and M. Quigley, "Light Drinking in Pregnancy, a Risk for Behavioural Problems and Cognitive Deficits at 3 Years of Age?" *International Journal of Epidemiology* 38, no. 1 (2009):129–140.

11. Association of Professors of Gynecology and Obstetrics, *Nausea and Vomiting of Pregnancy* (Washington, DC: Association of Professors of Gynecology and Obstetrics, 2001); T. M. Goodwin, "Hyperemesis Gravidarum," *Obstetrics & Gynecology Clinics of North America* 35 (2008):401.

12. Association of Professors of Gynecology and Obstetrics, *Nausea and Vomiting of Pregnancy;* S. M. Flaxman and P. W. Sherman, "Morning Sickness: A Mechanism for Protecting Mother and Embryo," *Quarterly Review of Biology* 75 (2000):113.

13. R. E. Gilbert, "Epidemiology of Infection in Pregnant Women," in *Congenital Toxoplasmosis: Scientific Background, Clinical Management and Control,* 1st ed., edited by E. Petersen and P. Amboise-Thomas (Paris: Springer-Verlag, 200).

14. H. Osman, I. M. Usta, N. Rubeiz, et al., "Cocoa Butter Lotion for Prevention of Striae Gravidarum: A Double-Blind, Randomised and Placebo-Controlled Trial," *BJOG* 115 (2008):1138.

15. G. L. Young and D. Jewell, "Creams for Preventing Stretch Marks in Pregnancy," *Cochrane Database of Systematic Reviews* (2000):CD00066.

16. M. Beckmann and A. Garrett, "Antenatal Perineal Massage for Reducing Perineal Trauma," *Cochrane Database of Systematic Reviews* (2006): CD005123.

10. CHILDBIRTH

1. B. A. Bucklin, J. L. Hawkins, J. R. Anderson, and F. A. Ullrich, "Obstetric Anesthesia Workforce Survey: Twenty-Year Update," *Anesthesiology* 103 (2005):645.

2. K. C. Johnson and B. A. Daviss, "Outcomes of Planned Home Births with

Certified Professional Midwives: Large Prospective Study in North America," *BMJ* 330 (2005):1416.

3. O. Olsen, "Meta-Analysis of the Safety of Home Birth," *Birth* 24 (1997):4.

4. E. R. Cluett and E. Burns, "Immersion in Water in Labour and Birth," *Cochrane Database of Systematic Reviews* (2009):CD000111.

5. Agency for Healthcare Research and Quality, Evidence Report/Technology Assessment number 133, *Cesarean Delivery on Maternal Request*, March 2006, available at www.ahrq.gov/clinic/tp/cesarreqtp.htm#Report (accessed December 5, 2).

6. M. F. MacDorman, E. Declercq, F. Menacker, and M. H. Malloy, "Neonatal Mortality for Primary Cesarean and Vaginal Births to Low-Risk Women: Application of an 'Intention-to-Treat' Model," *Birth* 35 (2008):3.

7. E. A. Frankman, L. Wang, C. H. Bunker, and J. L. Lowder, "Episiotomy in the United States: Has Anything Changed?" *American Journal of Obstetrics and Gynecology* 200 (2009):573.

8. G. Carroli and J. Belizan, "Episiotomy for Vaginal Birth," *Cochrane Database of Systematic Reviews* (2000):CD000081; K. Hartmann, M. Viswanathan, R. Palmieri, G. Gertlehener, J. Thorp, and K. N. Lohr, "Outcomes of Routine Episiotomy: A Systematic Review," *JAMA* 293 (2005):2141.

9. R. C. Pattinson, "Pelvimetry for Fetal Cephalic Presentations at Term," *Cochrane Database of Systematic Reviews* (2000):CD000161.

10. H. Rabe, G. Reynolds, and J. Diaz-Rossello, "Early Versus Delayed Umbilical Cord Clamping in Preterm Infants," *Cochrane Database of Systematic Reviews* (2004):CD003248; A. C. Yao and J. Lind, "Effect of Gravity on Placental Transfusion," *Lancet* 2 (1969):505; J. M. Ceriani Cernadas, G. Carroli, L. Pellegrini, L. Otano, M. Ferreira, C. Ricci, O. Casas, D. Giordano, and J. Lardizabal, "The Effect of Timing of Cord Clamping on Neonatal Venous Hematocrit Values and Clinical Outcome at Term: A Randomized, Controlled Trial," *Pediatrics* 117 (2006):779; R. G. Strauss, D. M. Mock, K. J. Johnson, G. A. Cress, L. F. Burmeister, M. B. Zimmerman, E. F. Bell, and A. Rijhsinghani, "A Randomized Clinical Trial Comparing Immediate Versus Delayed Clamping of the Umbilical Cord in Preterm Infants: Short-Term Clinical and Laboratory Endpoints," *Transfusion* 48 (2008):658; J. S. Mercer, B. R. Vohr, M. M. McGrath, J. F. Padbury, M. Wallach, and W. Oh, "Delayed Cord Clamping in Very Preterm Infants Reduces the Incidence of Intraventricular Hemorrhage and Late-Onset Sepsis: A Randomized, Controlled Trial," *Pediatrics* 117 (2006):1235.

11. M. Steiner, "Postpartum Psychiatric Disorders," *Canadian Journal of Psychiatry* 35 (1990):89.

11. MENOPAUSE

1. "The Role of Testosterone Therapy in Postmenopausal Women: Position Statement of the North American Menopause Society," *Menopause* 12 (2005):496.

2. J. E. Rossouw, G. L. Anderson, R. L. Prentice, et al., "Risks and Benefits of Estrogen Plus Progestin in Healthy Postmenopausal Women: Principal Results from the Women's Health Initiative Randomized Controlled Trial," *JAMA* 288 (2002):321.

12. BOOBS

1. K. Ashizawa, A. Sugane, and T. Gunji, "Breast Form Changes Resulting from a Certain Brassière," *Journal of Human Ergology* (Tokyo) 19 (June 1990):53–62. http://www.e-sante.be/soutien-gorge-question/sport-sante-77-243-6294.htm.
2. Susan Love, Karen Lindsey, and Marcia Williams, *Dr. Susan Love's Breast Book,* 3d rev. ed. (New York: HarperCollins Publishers, 2000.)
3.
4. *Pediatric Surgery Update* 26, no. 01 (January 2006).
5. National Institutes of Health Consensus Development Conference Statement Jan 21–23, 1997, 103, "Breast Cancer Screening for Women Ages 40–49," www.consensus.nih.gov/cons/103/103_intro.htm.
6. M. Elwood, B. Cox, and A. Richardson, "The Effectiveness of Breast Cancer Screening by Mammography in Younger Women: Correction," Online *Journal of Current Clinical Trials* 121 (1994):385.
7. L. L. Humphrey, M. Helfand, B. K. Chan, and S. H. Woolf, "Breast Cancer Screening: A Summary of the Evidence for the U.S. Preventive Services Task Force, *Annals of Internal Medicine* 137 (2002):347.
8. K. Armstrong, E. Moye, S. Williams, J. A. Berlin, and E. E. Reynolds, "Screening Mammography in Women 40 to 49 Years of Age: A Systematic Review for the American College of Physicians," *Annals of Internal Medicine* 146 (2007):516.
9. L. L. Humphrey, M. Helfand, B. K. Chan, and S. H. Woolf, "Breast Cancer Screening: A Summary of the Evidence for the U.S. Preventive Services Task Force," *Annals of Internal Medicine* 137 (2002):347.
10. A. B. Miller, T. To, C. J. Baines, and C. Wall, "The Canadian National Breast Cancer Screening Study-1: Breast Cancer Mortality After 11 to 16 Years of Follow-Up A Randomized Screening Trial of Mammography in Women Age 40 to 49 of Age," *Annals of Internal Medicine* 137 (2002):305–312.
11. L. A. G. Ries, D. Melbert, M. Krapcho, A. Mariotto, B. A. Miller, E. J. Feuer, L. Clegg, M. J. Horner, N. Howlader, M. P. Eisner, M. Reichman, and B. K. Edwards (eds.), *SEER Cancer Statistics Review, 1975–2004* (Bethesda, MD: National Cancer Institute).
12. http://seer.cancer.gov/csr/1975_2004/results_single/sect_04_table.14_2pgs.pdf.
13. P. D. Darbre, "Aluminium, Antiperspirants and Breast Cancer," *Journal of Inorganic Biochemistry* 99, no. 9 (2005):1912–1919.
14. P. W. Harvey and D. J. Everett, "Significance of the Detection of Esters of

P-hydroxybenzoic Acid (Parabens) in Human Breast Tumours," *Journal of Applied Toxicology* 24, no. 1 (2004):1–4.

15. D. K. Mirick, S. Davis, and D. B. Thomas, "Antiperspirant Use and the Risk of Breast Cancer," *Journal of the National Cancer Institute* 94, no. 20 (2002):1578–1580; S. Fakri, A. Al-Azzawi, and N. Al-Tawil, "Antiperspirant Use as a Risk Factor for Breast Cancer in Iraq," *Eastern Mediterranean Health Journal* 12, nos. 3–4 (2006):478–482.

16. K. G. McGrath, "An Earlier Age of Breast Cancer Diagnosis Related to More Frequent Use of Antiperspirants/Deodorants and Underarm Shaving," *European Journal of Cancer* 12, no. 6 (2003):479–485.

13. PEE

1. E. J. Hay-Smith, K. Bo, L. Berghmans, et al. "Pelvic Floor Muscle Training for Urinary Incontinence in Women (Cochrane Review)," *Cochrane Database of Systematic Reviews* 1 (2001):CD001407; T. A. Shamliyan, R. L. Kane, J. Wyman, and T. J. Wilt, "Systematic Review: Randomized, Controlled Trials of Nonsurgical Treatments for Urinary Incontinence in Women," *Annals of Internal Medicine* 148 (2008):459.

14. BUTTS

1. A. J. Miles, T. G. Allen-Mersh, and C. Wastell, Department of Surgery, Westminster Hospital, London, "Effect of Anoreceptive Intercourse on Anorectal Function," *Journal of the Royal Society of Medicine* 86, no. 3 (1993):144–147.

2. A. C. Kinsey, W. B. Pomeroy, and C. E. Martin, *Sexual Behavior in the Human Male* (Philadelphia, PA: W. B. Saunders, 1948).

3. Ung-hoe Yi, Jong-seong Sin, and Hyeong-gi Choe, "Sexual Behavior of Korean Women," *Daehan Namseong Gwahak Hoeji* 17, no. 3 (1999):177–185.

4. N. J. Talley and M. Jones, "Self-Reported Rectal Bleeding in a United States Community: Prevalence, Risk Factors, and Health Care Seeking," *American Journal of Gastroentrology* 93 (1998):2179.